EarthScore

Your Personal Environmental Audit & Guide

Donald W. Lotter

MORNING SUN PRESS

ISBN 0-9629069-4-8

Publisher: Jack Howell
Morning Sun Press
P.O. Box 413
Lafayette, CA 94549
(510) 932-1383

Book and Cover Design:
Teutschel Design Services
Palo Alto, CA 94306

Library of Congress Catalogue-in-Publication Data

Lotter, Donald, 1952-
Earthscore : a personal environmental audit & guide / Donald Lotter.
p. cm.
Includes bibliographical references.
ISBN 0-9629069-4-8 : $8.95
1. Environmental responsibility–Evaluation. 2. Environmental responsibility–Handbooks, manuals, etc. 3. Man–Influence on nature–Evaluation. 4. Man–Influence on nature–Handbooks, manuals, etc. I. Title.
GE195.7.L68 1993
363.7'0525–dc20 93-30492
CIP

Printed on recycled, acid-free paper
Printed in the United States of America

Individual copies: $8.95 plus $1.75 for postage and handling.
Discounts available for quantity orders.

Morning Sun Press books are available at quantity discounts when used to promote products or services. For more information, please write to Premium Marketing Department, Morning Sun Press, P.O. Box 413, Lafayette, CA 94549 (510) 932-1383.

10 9 8 7 6 5 4 3 2 1

EarthScore

Your
Personal
Environmental
Audit & Guide

Donald W. Lotter

Table of Contents

Acknowledgments vii

Foreword ix

Introduction xi

Section 1
Household Energy: General 1

Section 2
Household Energy: Winter 7

Section 3
Household Energy: Summer 17

Section 4
Water 23

Section 5
Transportation 31

Section 6
Consumerism: Durable Goods 41

Section 7
Consumerism: Food and Agricultural Products 47

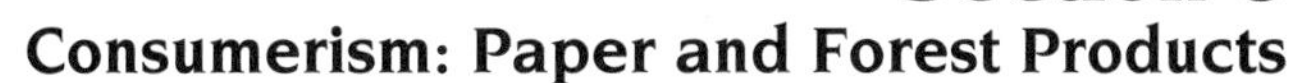

Section 8
Consumerism: Paper and Forest Products 55

Section 9
Toxics 65

Section 10
Waste, Packaging, Single-Use Items, and Recycling 75

Section 11
Environmental Advocacy 85

Section 12
Respect for the Land 91

Section 13
Livelihood 95

Section 14
Family Planning 99

All-Star Bibliography 103

EarthScore Totals 104

EarthScore Chart 105

Acknowledgments

Some of the people I am indebted to for their help in developing and for their review of *EarthScore* are Elaine Herbert of the Yolo Environmental Resource Center, California; Bill Knox of the Energy Management Services of Davis, California; Mark Eisen of Home Depot; Pamela Brown of the City of Portland Environmental Commission, Oregon; Marilyn Miller of the City of Ojai Recycling Program, California; Joel Makower of the *Green Consumer Letter*; Richard Kahlenberg; and Jeanne Trombly. I also thank once again all of the people who helped with the *EnviroAccount* software through the years of its development. Most importantly, I thank my parents, Will and Jane Lotter, for their faithful support throughout the *EnviroAccount/EarthScore* project.

Foreword

For years, I've told my readers and audiences, "Every time you open your wallet, you cast a vote for or against the environment." What's excited me most about this notion is that the marketplace is not a democracy: It doesn't take 51 percent of the population voting one way or another to effect change. Far from it. Just a small percentage of people changing the way they shop and the way they live can be a powerful force, persuading politicians, moving markets, and fomenting fellow citizens to change.

But things aren't always that simple.

Those of us who make it our business to understand the relationships between the things we buy and do and their impact on the environment know well the challenges and confusion each of us faces in trying to "go green." The challenges come when you try to do the right things, not simply perform a few symbolic acts every third week in April around Earth Day.

The confusion sets in when you try to truly grasp the environmental implications of your purchases and actions. Understanding the envirormental impacts of something as simple as using a polystyrene foam coffee cup instead of a paper cup or a washable mug requires digesting an impressive amount of technical information. Much of that information is incomplete, is inaccurate, or conflicts with other data. Never mind addressing the real issues of the day–how to minimize the impacts of commuting to work, managing our cities, and cooling and heating our homes in environmentally responsible ways.

The result is that we often do the wrong things, or do nothing at all. And amid the frustration and confusion, it's hard to maintain perspective. Consider this: Even if you were to make all the "right" environmental choices during your next trip to the supermarket, the benefits to the earth would be wiped out by the mere act of driving 1 mile to and from the store in a poorly tuned, gas-guzzling car with a cold engine and underinflated tires. The point is, it's crucial that we undertake the significant acts as well as the symbolic ones.

The truth is, none of us will ever get everything completely right. It's simply not possible. Every day, even the "greenest" of us uses resources and creates wastes that must be disposed of. That makes each of us a contributor, however unwittingly, to environmental degradation. The most we can hope for is to do the least amount of damage in the pursuit of life, liberty, and happiness.

But that doesn't mean your actions can't have a powerful impact on the environment or be a potent force in persuading others to be more responsible. All it takes is a little knowledge.

That's where this book comes in.

What's great about *EarthScore* is its ability to help you get a handle on the way you live and the impact of your actions on the earth. That puts this book head and shoulders above most other green guides, which lump dozens of disparate lifestyles into one gigantic pot called "society." Ultimately, these books try to reach everyone with the same set of advice. They're kind of like the chair designed for the "average" person: It tends to be too big for half the population and too small for the rest.

EarthScore's innovative and personalized scoring system means that the book can be a perfect fit for every reader. By understanding and tracking your own habits and lifestyles over time, you can identify where you can truly make a difference. That will be different for every person and every household. And over time, as each of us grows and changes, our impacts on the earth change, too. We use different types and amounts of resources and create different types and amounts of waste as we set up a household, find a mate, perhaps raise kids, change jobs, change homes, take vacations, retire, and on and on. With *EarthScore*, you can track these changes–and their impacts–and decide what if any alterations may be appropriate.

In the end, *EarthScore* is about self-knowledge. Without self-knowledge, meaningful change is simply impossible, and our best intentions to help the earth will be for naught.

Joel Makower,
Author, *The Green Consumer*; Editor, *The Green Consumer Letter*

Introduction

Your Connections to the Earth

This book will connect you to the earth. To be more precise, it will help make you aware of your connections to the earth. Physically, we interact with the earth in hundreds of different ways, most often indirectly through the things we buy–like food, consumer goods, and electricity–but also directly in acts such as driving and gardening. Becoming aware of our individual links to the biosphere is a first, necessary step toward saving the earth for future generations. Remaining aware of these complex connections without ever seeing them in graphical form is difficult. *EarthScore* is designed to help you visualize your environmental connections.

EarthScore is based on a computer software program that I developed called *EnviroAccount* (see the back page of this book). The book is designed to organize and present to you the various magnitudes of each of your major links to the earth. It is in the magnitudes of each link that you will be able to see progress or regress over time (you would hope for progress, but seeing regress is no reason to stop using *EarthScore*; more on this follows). The process of reducing your impact on the environment in the most efficient way will be one of minimizing the magnitudes of various flows, or throughputs: different forms of energy, water, products, and so on. Your task is to optimize these magnitudes and to balance your living needs–food, shelter, livelihood, recreation, procreation–with your environmental impact goals. *EarthScore* helps you to set priorities and determine the areas where you can most effectively reduce your environmental impacts.

Awareness versus Guilt

Everyone should be interested in becoming aware of his or her environmental connections, even the individual who has little interest in or time to invest in the environment. For the majority of people who have been relatively honest with themselves and in their livelihood, taking this audit should not be an experience of guilt or shame. Awareness, even without the ability or willingness to change, is important, because the process of becoming aware allows the seed for change to be planted. The change could come 5 years down the road, or it could happen surprisingly rapidly. So even if you have a high-impact lifestyle, complete *EarthScore* just for fun, and don't feel bad if you can't do much about your impact score at this time. Keep an awareness of your various impact levels in the back of your mind, and someday you may get the opportunity to make that significant and long-mulled-over change.

How *EarthScore* Works: Impact and Action Points

EarthScore consists of 107 questions in 14 sections. There are two types of questions in *EarthScore*: Impact Questions, which give Impact Points, and Action Questions, which give Action Points. Impact Points measure your environmental impact, (how much you harm the earth), while Action Points measure the good things you do for the earth.

At the end of each section, you total your Impact and Action Points separately, and then, at the end of the book, you add all of the section totals to get a grand total for each category. You will then be able to determine your Impact Rating, ranging from Eco-Titan to Eco-*Tyrannosaurus rex*.

EarthScore is meant to be completed at least twice, the second time 6 months to 1 year after the first, so that you can see any changes. You can reuse your copy of *EarthScore* and its chart by writing with a different-color pencil the second time.

After summing up the section totals, you fill out the *EarthScore* chart,which gives you a graphical display of the magnitudes of your various impacts and actions. This helps you to visualize your connections and set goals for yourself or your family. To facilitate the development of your personal environmental awareness, you can put up your *EarthScore* chart someplace where you can see it occasionally.

Impact Points approximate your impact on the environment based on three criteria: (1) pollution, (2) use of nonrenewable resources, and (3) degradation of ecological systems. As mentioned, previously these impacts can occur both directly or indirectly from the various things you do.

The magnitudes of Impact and Action Points are based on the formulas used in the *EnviroAccount* software. These were developed using environmental information from current books and journals and were further refined based on feedback from the users of *EnviroAccount*. Impact and Action Points are not meant to be accurate and precise, as accuracy and precision do not yet exist in the newly developing field of impact assessment. For this reason, categories throughout *EarthScore* are often not mutually exclusive (that is, $15-$30, followed by $30-$60 rather than $15.00-$29.99, $30.00-$60.00 and so on). Such conciseness would falsely impart an impression of precision. All quantities in *EarthScore*, including scores, should be considered approximate. The more important thing is to get a rough quantitative handle on your relative environmental impacts so that you can monitor your progress or compare yourself with someone with another lifestyle.

A good example of the comparison of lifestyles is between country dwellers and city dwellers. City dwellers (not suburban dwellers) often think of themselves as being environmentally "bad" because they live in the "polluted" city, while country dwellers generally feel the opposite because they live in a more pristine environment. Upon analysis using *EnviroAccount* or *EarthScore*, however, the city dwellers often have a significantly *lower* environmental impact, mostly because country dwellers generally drive great distances to commute into the cities and because city dwellers often have more compact habitations.

For a more accurate assessment of your environmental impacts and actions, and a more sophisticated presentation, you may want to obtain a copy of the *EnviroAccount* software, as detailed on the last page of this book.

Information and Resources

Following most questions is information about why the question is important, what you can do to improve, and resources such as books and hotlines. *EarthScore* emphasizes phone numbers because the telephone provides the best means of taking immediate action.
The six publications listed in the All-Star Bibliography in the back of *EarthScore* have been chosen from hundreds of books as the best, most up-to-date, most comprehensive, most respected resources for becoming environmentally informed. Five of these books cover different and important areas: consumerism and transportation (*Green Consumer*), home design and improvement (*Efficient House Sourcebook*), food (*Recipes from an Ecological Kitchen*), money and livelihood (*Your Money or Your Life*), and the natural world (*Diversity of Life*); a general directory (*Stewart's Environmental Directory*). I cite two of these books, *Green Consumer* and *Efficient House Sourcebook*, as resources for many individual topics, and I recommend both of them highly. I have cited dozens of other books under specific topics for those who are interested in pursuing those particular subjects.

Have fun!

Don Lotter
Davis, California

Section 1

Household Energy: General

The first three sections of *EarthScore* concern household energy: general uses, winter heating, and summer cooling. After the industrial and transportation sectors of the United States economy, our homes take the third largest amount of energy. By far the greatest share of household energy goes toward heating. After heating, the ten other big energy users in the home are the

1. water heater (16%)
2. refrigerator/freezer (12%)
3. air conditioner (8%-30%)
4. range (6%)
5. lights (7%)
6. waterbed
7. color television
8. washer
9. dryer
10. space heater

Section 1

Household Energy Use: General

Circle an answer in each question. Record your points in the box provided.

IMPACT **ACTION**

1.1

My average monthly energy bill, per person, is approximately _____.

	POINTS
1. less than $15	6
2. $15 - $30	12
3. $30 - $60	18
4. $60 - $90	24
5. $90 - $120	30
6. $120 - $150	40
7. $150 - $180	50
8. more than $180	60

Home energy use can be reduced by an estimated 40% to 70% through proper maintenance and retrofitting. This means that your energy bill would be reduced by that same amount. The rest of the questions in this section cover the major areas in which your can make these savings.

Resources

Makower, *The Green Consumer*. Home energy chapter. See the All-Star Bibliography.

Sardinsky, *Efficient House Sourcebook*. See the All-Star Bibliography.

1.2

I have invested _____ effort/money in installing solar energy equipment around my home for reducing my dependence on energy from nonrenewable sources.

1. no	0
2. a little	3
3. moderate	6
4. fairly extensive	9
5. a great deal of	12

IMPACT | ACTION

Resources

John Schaeffer, *Alternative Energy Sourcebook*,7th edition. Ukiah, CA: Real Goods Trading Corporation, 1993. 1-800-762-7325. $16.One of the most entertaining house-improvement books to browse through.An excellent source of energy sensible technologies,chock-full of pictures and tips along with energy products for sale.

The American Solar Energy Society. 303-443-3130.
Sardinsky, *Efficient House Sourcebook*. See All-Star Bibliography.

1.3

I have invested _____ in energy-saving appliances for my home.

1. none/very little	0
2. a little	3
3. moderately	6
4. fairly extensively	9
5. a great deal	12

Generally, the most important appliance for energy savings is the refrigerator. You can save energy in the kitchen in many other ways. For instance, if you often use an oven to heat or cook small or medium-sized items, you can save 50% of the energy by using a microwave.

Resources

Makower, *The Green Consumer*. See All-Star Bibliography.

Sardinsky, *Efficient House Sourcebook*. See All-Star Bibliography.

1.4

I have installed energy-saving compact fluorescent lights in _____ of my commonly used light fixtures.

1. 3/4	1
2. 1/2	2
3. 1/4	3
4. none	4

The electricity used by a conventional light bulb costs about five to ten times the price of the bulb. For this reason it is worth the money to buy compact fluorescent (CF) bulbs, which last ten times longer (7,500 hours) and use one-fourth the energy (per hour) of an incandescent bulb, but generally cost about $18 to $20. Compact fluorescents are now being produced in many convenient sizes. If you need a conventional bulb, halogen bulbs are the most energy efficient.

IMPACT | ACTION

Conventional incandescent light bulbs are best used in areas where you only turn on the light for brief periods of time.

Resources

Makower, *The Green Consumer*. See All-Star Bibliography. This has a full listing of CF light bulbs.

Sardinsky, *Efficient House Sourcebook*. See All-Star Bibliography.

1.5

My water heater is _____.

1. solar 1
2. gas and insulated 3
3. gas and uninsulated 6
4. electric and insulated 9
5. electric and uninsulated 12

Water heating is generally the home's second biggest energy use, after house heating, accounting for about 16% of total energy. Heating water is significantly cheaper with gas than with electricity. If your water heater is uninsulated you can purchase a water heater blanket from a hardware store. Many water heaters have built-in insulation. If the tank feels warm on a cool day, it needs an insulating blanket. You should also insulate the pipes leading from the tank, if possible. Your water heater thermostat should be turned to 130°F for optimum efficiency. If your water heater pings as it warms up water, it needs to be drained of sediments.

1.6

I have an energy-saver refrigerator. I keep it at optimum temperature (38-40°F), and I unplug the extra refrigerator if it is not fully used.

1. All of the above are true. 1
2. Two of the above are true. 2
3. One of the above are true. 3
4. None of the above are true. 4

If you live in a city, your refrigerator probably uses 25% of your electricity. Energy-efficient appliances generally use 50% less energy than old appliances and often pay for themselves in just a few years. Keeping the coils at the back of the refrigerator clean also saves energy.

IMPACT | ACTION

> **Resources**
>
> Makower, *The Green Consumer* chapter about energy efficient appliances. See All-Star Bibliography.
>
> Sardinsky, *Efficient House Sourcebook*. See All-Star Bibliography.

1.7

During sunny weather, I ______ dry my clothes on a clothesline.

1. never	0
2. occasionally	1
3. frequently	2
4. usually	3
5. always	4

On a warm, sunny day, producing heat electrically to dry clothes is a waste of energy.

Scores

Total the points for this section:

IMPACT | ACTION

Section 2

Household Energy: Winter

If you want to reduce your home energy use, winter heating and heat conservation are the first places to look. Up to 60% of total household energy goes to heating the home. This section evaluates your efforts to conserve heat energy plus gives information and resources on how to improve.

At the end of this section you will rate the winter coldness for your area from very cold to warm and then reduce or increase your points accordingly.

Section 2
Household Energy Use: Winter

Circle an answer in each question. Record your points in the box provided.

IMPACT | ACTION

2.1

In winter, during waking hours, I keep the house temperature around _____°F.

1. below 64 or off — 2
2. 64-66 — 4
3. 66-68 — 6
4. 68-70 — 8
5. 70 or above — 10

Pile-lined clothing is great for keeping warm around the house and saving energy. Pile fleece is light and comfortable and can pay for itself in one winter's energy savings. Sporting goods stores and mountaineering supply catalogs carry pile clothing.

Resources

Campmor. 1-800-526-4784. Sells outdoor equipment, including pile clothing.

2.2

In winter, during sleeping hours, I keep the house temperature around _____°F.

1. below 55 or off — 2
2. 55-59 — 4
3. 59-62 — 6
4. 62-66 — 8
5. 66 or above — 10

Do not sleep under an electric blanket. Research indicates that electric blankets can be a health hazard and can disrupt the immune system of those sleeping directly under them when they are turned on, due to the electromagnetic field. Children and pregnant women

IMPACT | ACTION

especially should avoid electric blankets. It is all right to warm a bed with an electric blanket before you get into it, but turn it off when you enter the bed.

Some utility companies will come to your house to do readings of electromagnetic fields.

Resources

Paul Brodeur, *The Currents of Death*. New York, NY: Simon and Schuster, 1989. Explains in detail the problem of electromagnetic fields and their effects on health.

2.3

The size of my living space is approximately ______. (Divide your total living space by the number of people living in it.)

1. 300 sq. ft. or less (17 feet by 17 feet) 3
2. 300 - 600 sq. ft. 6
3. 600 - 900 sq. ft. 9
4. 900 - 1500 sq. ft. 12
5. more than 1500 sq. ft. 15

For many people, cohousing and shared living are important ways to reduce environmental impacts and improve social life. Shared living means using less energy, space and "things" (such as vacuum cleaners, freezers, and toys) per person. Shared living can mean almost any level of sharing: multiple people sharing a house, multiple houses sharing yard space (a cluster), multiple clusters sharing goods and services. Shared housing, cooperative communities, cohousing, and village clusters are all approaches currently being practiced and rapidly being developed.

Resources

Shared Living Resource Center, 2375 Shattuck Ave., Berkeley, CA, 94704. 510-548-6608.

The Urban Ecologist. Berkeley, CA. Urban Ecology Institute. $25/yr. 510-843-0460.

Kathryn McCamant and Charles Durrett, *Cohousing: A Contemporary Approach to Housing Ourselves*. Berkeley, CA: Habitat Press, 1988. 510-843-0460. $22.

IMPACT	ACTION

2.4

The house I live in is insulated _____.

1.	super well (walls, ceiling, double paned windows)	2
2.	well	4
3.	moderately well	6
4.	only fairly well	8
5.	poorly	10

Those who use little or no heating in winter should answer 1 or 2.

If you plan to insulate, inform yourself about foam insulation. Foam insulation is often made from chlorofluorocarbons (CFCs) or other ozone-depleting chemicals. Other types of insulation, such as spun glass or cellulose, may be more appropriate.

If you have an attic, be sure that it is insulated.

Twelve times as much heat escapes from a single-pane window as through a wall. Drapes can reduce the heat lost through a window by 50% if the drapes are insulated and sealed at the bottom. Double-paned windows and reflective coatings are also effective at cutting heat loss.

Resources

Conservation and Renewable Energy Inquiry and Referral Service (CAREIRS). 1-800-523-2929.

Sardinsky, *Efficient House Sourcebook*. See All-Star Bibliography.

Harry Yost, *Home Insulation: Do It Yourself & Save As Much As 40 percent*. Pownal, VT: Storey Communications, 1991. $11.95

2.5

The doors, windows, pipes and electrical outlets in my house are caulked, weather-stripped, or sealed ______.

1	super well	1
2.	well	2
3.	adequately	3
4.	poorly	4

The cracks and gaps in the average U.S. house are equivalent to a 3-foot-by-3-foot hole in the wall. This translates to about 15% of home heating energy. These holes can be sealed with caulk. Doors and window frames can be sealed with specially designed weather-strip-

IMPACT **ACTION**

ping, available at hardware stores. Check for leaks on cold nights when the house is warm by feeling for cold air coming in.

According to experts, 99% of all houses in the U.S. with central furnaces or air conditioners have duct leaks.

Balance weather-stripping and sealing of your house with adequate ventilation to prevent the buildup of gases emanating from foam insulation, carpets, particleboard, and so on. In some cases, indoor air can be the source of 90% of the air pollution we breathe. A heat-exchange system may be recommended.

Resources

CAREIRS. 1-800-523-2929. Information about heat exchange systems.

John Bower, *The Healthy House*. New York, NY: Carol Publishing Group, 1991. 201-866-0490. $16.95. One of the most comprehensive books on indoor air pollution.

Sardinsky, *Efficient House Sourcebook*. See All-Star Bibliography.

The Environmental Protection Agency Indoor Air Quality Hotline. 1-800-438-4318.

The American Lung Association. 215-315-8700. Pamphlets on indoor air pollution.

2.6

I have had the following type of energy audit done on my house______.

1.	no audit	0
2.	informal self-audit	3
3.	self-audit with published guide	6
4.	free utility company audit	9
5.	professional audit	12

Call your utility company about an audit or referrals to professional home energy auditors.

IMPACT | ACTION

2.7

I have invested ______ time and/or money in increasing energy efficiency and decreasing my use of nonrenewable energy for heating my home.

(Question 1.3 in the previous section concerned how much you invest in energy efficient appliances. This question regards energy-efficient heating systems.)

1.	no/very little	0
2.	some	3
3.	moderate	6
4.	fairly extensive	9
5.	extensive	12

The average retrofit project pays for itself in four years of utility bill savings.

Ceiling fans circulate warm air from the ceiling, which can be 15°F warmer than air at the floor. A ceiling fan takes about the same amount of energy as a light bulb.

Passive solar heat gain occurs when the sun warms your house, and that heat is stored for use in the evening. See Resources for more information.

Furnaces need to be "tuned up" occasionally. This can be done by heat technicians for about $50.

Resources

CAREIRS. 1-800-523-2929.

National Audubon Society, *Building an Environmentally Friendly House*. Lincoln, MA: National Audubon Society Educational Resources Office. 617-259-9500. $3.75. A 40-page guide.

Sardinsky, *Efficient House Sourcebook*. See All-Star Bibliography.

Makower, *The Green Consumer*. See All-Star Bibliography.

IMPACT | ACTION

2.8

I burn __________ cords of firewood per year. (A cord is a pickup truck load.)

1. 0	0
2. less than 1/2	2
3. 1/2-1	4
4. 1-2	6
5. 2-3	8
6. more than 3	10

If you have an Environmental Protection Agency (EPA) certified catalytic stove, divide points by 3. The burning of wood can be a serious source of local air pollution. The points given here are for this pollution, not for energy use. A catalytic stove actually reburns the smoke particles and emits a fraction of the pollution of a regular wood burning stove.

Resources

For a list of EPA-certified stoves, call this EPA hotline: 703-308-8688.

Lawrence Tasaday, *Shopping for a Better Planet* New York: Meadowbrook Press, 1991. $10. Includes a section on catalytic stoves.

Scores

Total the points for this section:

IMPACT | ACTION

Now choose a winter coldness rating for your area from the following categories (these are multipliers, not points).

- 2.0 Very cold winters (average minimum temperatures below, 10°F)
- 1.5 Cold winters (average minimum temperatures between 10°F and 20°F)
- 1.0 Average winters (average minimum temperatures between 20°F and 30°F)
- 0.75 Cool winters (average minimum temperatures between 30°F and 40°F)
- 0.50 Warm winters (average minimum temperatures above 40°F)

IMPACT	ACTION

Enter your total number of Impact Points: ______

Enter your winter coldness rating: ______

Now multiply to get **adjusted Impact Points:** ______

Many other factors go into the environmental impact of home energy, such as how much nuclear vs. coal vs. hydropower is used in your region. If you want a more accurate assessment, you may want to consider the *EnviroAccount* software program (see back page).

Section 3

Household Energy: Summer

Electricity for air conditioning, the major home energy use during summer, is especially environmentally costly because demand comes during the afternoon at a time when electric power plants are overloaded. During these "peak hours" utility companies must start up oil burning plants to provide the extra electricity. This section evaluates your use of air conditioning and your efforts to minimize the use of energy for cooling your house.

Section 3
Household Energy Use: Summer

Circle an answer in each question. Record your points in the box provided.

IMPACT **ACTION**

3.1
On summer days, I use air-conditioning _____.

1.	never	0
2.	infrequently (10-20 days/yr.)	5
3.	fairly frequently (20-30 days/yr.)	10
4.	frequently (30-50 days/yr.)	15
5.	just about every day	20

Air conditioners consume enormous amounts of electricity, most of it during expensive "peak" hours from 12 to 6 PM. A fan uses about one-tenth of the energy of an air conditioner.

There are many ways to cool a house naturally to delay or prevent the need for turning on the air conditioner. Open windows at night to cool the house, then close them in the morning to retain cool air. Strategically placed trees (to allow sunlight in winter and provide shade in summer) can go a long way toward keeping a house cool and comfortable. In nonhumid regions, such as the Western United States, swamp coolers, which blow water-cooled air, are effective, energy efficient, and underutilized. Do you set the thermostat as high as possible (78°F)? Is your air conditioner shaded? (It should be.) Do you turn off the air conditioner when you leave the house for more than a half hour?

Air conditioner coils should be clean and straight and the filters clean.

Resources

CAREIRS. 1-800-523-2929.

Air Conditioning and Refrigeration Institute. 703-524-8800

IMPACT | ACTION

3.2

On summer nights, I use air conditioning _____.

1 never	0
2. infrequently (10 - 20 nights/yr.)	5
3. fairly frequently (20-40 nights/yr.)	10
4. frequently (40-60 nights/yr.)	15
5. just about every night	20

3.3

The southwest facing windows of my house are _____ shaded by trees, awnings, and trellises to keep direct sunlight out of the house during summer.

1. not at all	0
2. slightly	1
3. half	2
4. mostly	3
5. thoroughly	4

You can plant trees that shade in summer and shed their leaves in winter to allow the passage of sunlight into the house.

3.4

During summer, I make _________ effort to increase my tolerance to heat by such strategies as regular moderate exercise, reducing sugar and fat in my diet, eating plenty of fresh fruit, and drinking lots of water.

1. no	2
2. a little	4
3. moderate	6
4. good	8
5. great	10

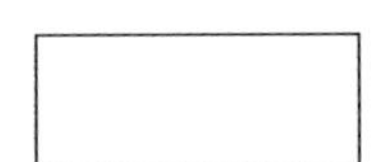

You will be surprised how well diet and exercise work to increase your tolerance for heat, if you can just get through the first 2 weeks. Start exercising easily.

IMPACT	ACTION

3.5

In summer, I reduce my use of electricity during peak hours of 12 to 6 P.M. _____.

1. always 1
2. most of the time 2
3. frequently 3
4. sometimes 4
5. rarely or never 5

When you use electricity during peak hours, along with many other people, utility companies must start up power plants that often must burn petroleum, which is both economically and environmentally costly. Peak hour rates can cost 4 times more per kilowatt hour than the lowest night rates.

3.6

In summer, I use a solar cooker to cook my food _____.

1. never 0
2. once a month or less 1
3. once a week 2
4. twice a week 3
5. more than twice a week 4

Solar cooking saves energy in two places: at the oven and at the air conditioner. Cooking with the sun helps to keep the house temperature down by eliminating the need to use the oven. Many people use solar ovens at least once a week, and the food they cook using this method is often better than anything from a conventional oven, perhaps because the cooking time is usually longer.

Resources

Beth Halacy and Dan Halacy, *Cooking With the Sun: How To Build and Use Solar Cookers*.Lafayette, CA: Morning Sun Press, 1992. 510-932-1383. $7.95.Gives a complete description of how to build an inexpensive solar oven, as well as a solar hot plate. The hot plate will saute foods normally prepared on stove top.

Solar Box Cookers International. 916-444-6616. Low priced solar ovens.

Real Goods Trading Company. 1-800-762-7325. Fancier, more expensive, and more durable solar ovens.

Scores

IMPACT	ACTION
IMPACT	ACTION

Total the points for this section:

Section 4

Water

Water is absolutely central to life. We are made of it, we cleanse ourselves with it, we play in it, and we use it in a thousand different ways. Yet despite its centrality to our existence, water is one of the most poorly managed of our resources.

Developing good water-use habits is of utmost importance, especially in the Western United States, where water looms as the major political and environmental issue of the future.

This section evaluates your water use and your efforts to conserve. At the end of this section, you will be asked to rate water scarcity for your region. You will then increase or reduce your water points accordingly. If you are in a region with plenty of water, Impact Points for water use will be minimal.

Section 4
Water

Circle an answer in each question. Record your points in the box provided.

IMPACT | ACTION

4.1

My watered lawn and garden space is _____. (Exclude food gardens.)

1.	less than 200 sq.ft.	2
2.	200-500 sq. ft.	4
3.	500-1,000 sq. ft.	8
4.	1,000-3,000 sq. ft.	16
5.	more than 3,000 sq. ft.	32

Lawns and gardens commonly account for 80% of the water used in a suburban household. A week of watering the above lawn sizes would take a minimum of 100, 200, 500, 1,000 and 2,000 gallons of water. Inefficient watering or hot spells can double these numbers. Watering in the early morning or at night reduces the amount of water that evaporates before water gets into the soil.

If you have water-saving grasses–buffalo grass, wheatgrass, zoysia grass, tall or fine fescue, or bermuda grass - reduce your score by one classification.

Resources

"Drought Survival Guide", Sunset, May 1991. Sunset Drought Guides, 80 Willow Rd., Menlo Park, CA, 94025. $2.50 for a reprint. An excellent 30 page guide to increasing water efficiency in your yard and garden. $7 for a set; set includes guides for drip irrigation, plants, and lawn watering.

IMPACT | ACTION

4.2

By such strategies as planting drought resistant lawn and garden plants (xeriscaping), using drip irrigation, planting trees and using other techniques, I have reduced my water consumption _____.

1. not at all	0
2. little	3
3. moderately	6
4. very significantly	9
5. greatly	12

Resources

AgAccess Bookstore. 916-756-7177. An international clearing house for books on all subjects related to agriculture, gardening, and land use. Ask for their water catalog.

4.3

I _____.

1. do not have a swimming pool	0
2. have a swimming pool that doesn't leak	4
3. have a swimming pool that leaks some	6
4. have a swimming pool that leaks a lot	8

About one in twenty pools leaks. Just a small leak in a pool can result in the loss of 700 gallons of water a day. If your pool loses more than a quarter inch of water per day (a half inch in hot, dry regions), you probably have a leak. A swimming pool loses no more water through evaporation than an average lawn of the same dimension.

To determine the number of gallons your pool loses per day (both evaporation and leakage), use one of the following formulas:

pool length (in feet) x pool width x inches of water lost per day x .6233

For round pools or spas:
ft. diameter x ft. diameter x inches of water lost per day x .48929

IMPACT | ACTION

4.4

I use _______ water to wash my car and other outside things like my patio.

1. no	0
2. a little	1
3. a moderate amount of	2
4. a lot of	3

A trigger nozzle on your hose will minimize waste of water.

4.5

I have installed low-flow shower heads and low-volume flush devices on my toilets.

1. None of the above is true.	0
2. One of the above is true.	1
3. Two of the above are true.	2
4. This is true for all of my showers and toilets.	3

Faucet aerators reduce flow by 50% and yet maintain good pressure. Low-flow shower heads reduce water use by 50% or more and, on the average, reduce water and electricity bills by 27¢ and 51¢ per day, respectively.

Resources

Earth Tools. 1-800-825-6460. Low-flow shower heads available.

Trademark Sales and Marketing. 414-727-1818. Low volume flush devices (that cut water volume by 55%) available.

4.6

Leaky faucets or pipes in my house or on my property cause _______ waste of water.

1. no or very little	0
2. a little	2
3. some	4
4. a fair amount of	6
5. extensive	8

A small leak from a faucet can waste 50 gallons of water per day. A leaky toilet can waste 8,000 gallons a month. To identify a toilet leak, put some dye in the tank. If it shows up in the bowl without flushing, you have a leak.

IMPACT	ACTION

4.7

I have invested _______ in water-saving technologies such as gray water systems and ultralow-flow toilets.

1. nothing or very little	0
2. a little	2
3. moderately	4
4. extensively	6
5. very extensively	8

A gray water system utilizes rinse water from the house for watering lawns and gardens. One of the niftiest is a basin that fits onto your toilet tank. Water that you wash with then becomes toilet flush water (gray water). See Real Goods Trading Company, under Resources.

An ultralow-flow toilet uses no more than 1.6 gallons of water per flush.

Resources

Resources Conservation. 1-800-243-2862. Water-saving devices available.

Robert Kourik, *Gray Water Use in the Landscape: How to Help Your Landscape Prosper with Recycled Water.* Santa Rosa, CA: Metamorphic Press, 1988.

Real Goods Trading Company. 1-800-762-7325. Gray water systems available.

4.8

I make _______ effort to conserve water by doing such things as minimizing my shower flow and flushing every other time I use the toilet.

1. no or very little	0
2. some	1
3. a moderately good	2
4. a really good	3
5. great	4

Resources

Randall Schultz, *Turn Off the Tap: How to Cut Your Water Usage by 50 Percent.* Albuquerque, NM: Creative Designs, 1991. 1-800-869-8520. $3.50.

IMPACT	ACTION
IMPACT	ACTION

Scores

Total the points for this section:

On a scale of 0.25 to 3.0 rate the water scarcity for your area and enter it below.

0.25	(water very plentiful all year)
0.5	(water fairly plentiful)
1.0	(some scarcity of water)
2.0	(water scarce)
3.0	(water very scarce)

Examples: Southern California = 2.5; San Francisco Bay Area = 2.0; Atlantic seaboard = 0.5 to 1.0

Enter your water scarcity rating: _____

Enter your Impact Point total for this section: _____

Now multiply these to get **adjusted Impact Points:** _____

IMPACT

Section 5
Transportation

Operating an automobile has a greater environmental impact than any other common human activity except for having too many children. Not only does it eat up fossil fuels at an enormous rate per person, it is also a major cause of greenhouse warming and other air pollution problems. (see the table in 5.10). Additionally, we pave over huge amounts of land for drivers. In urban areas, one-third to one-half of all land is given over to the automobile in some way: for streets, parking lots, gas stations, repair shops, junkyards, and so on. This section evaluates your transportation energy use and efforts to promote efficiency and to minimize pollution.

Section 5
Transportation

Circle an answer in each question. Record your points in the box provided.

IMPACT	ACTION

5.1

My main vehicle gets _____ miles per gallon of gasoline.

1. I don't own a car.	0
2. I have an electric car.	1
3. 45 or more	3
4. 35-45	6
5. 25-35	9
6. 15-25	12
7. less than 15	15

The world's known petroleum reserves will be used up in 35 years at the present rate of consumption, mostly by driving. The U.S., with only 5% of the world's population, uses 35% of the world's fossil-fuel energy.

Keeping your car well tuned is the best way to get good mileage. Other strategies are to keep tires properly inflated, keep your speed down, and refrain from warming up the engine when you start it.

5.2

I drive _______ miles per year. (If you share a car or cars with another person, then divide the total mileage by 2. Do not count miles driven at your work; do count commute mileage.)

1. 0-1,000	4
2. 1,000-4,000	8
3. 4,000-8,000	16
4. 8,000-12,000	24
5. 12,000-16,000	32
6. 16,000-20,000	40
7. 20,000-24,000	48
8. 24,000-30,000	56
9. more than 30,000	64

IMPACT | ACTION

Driving can be put on a highly subjective continuum between two poles: necessary and unnecessary. Unnecessary driving is a tough habit to break when fuel is as cheap as it is in the United States. People in this country tend to use driving as an escape–indeed almost as therapy. When you drive, you are alone with your thoughts and yet stimulated by the movement of the passing landscape and the stereo. For many, driving becomes an environmentally costly form of meditation. Breaking the habit of unnecessary driving is where spirituality and environment meet in their most immediate way. At first it hurts to deny yourself that 50-mile round-trip to a "happening" but unnecessary event and instead to stay home or visit a nearby friend. There is no clear demarcation between unnecessary and necessary driving; the line is a subjective, personal matter, ever shifting. "It's boring to stay here," you hear yourself and others say when you deny yourself that trip. Delve into the boredom and explore the spiritual potential in it.

5.3

The engine in my vehicle is in _______ condition.

1. new or newly rebuilt 2
2. good 4
3. fair 6
4. poor 8
5. bad 10

Your car's emissions will be much higher if its engine is old and in poor condition.

5.4

I have my car tuned ______.

1. as recommended in the owner's manual 0
2. when it is hard to start 10
3. when it breaks down 20

By far the most important criterion for keeping emissions down (and getting good mileage) is whether the car's engine is in good condition and kept well tuned. According to the Environmental Protection Agency, if you have an older car, keeping it well tuned can reduce pollution by 40%.

IMPACT | ACTION

5.5

I take an average of ______ short distance car trips (less than 2 miles) per week.

1. less than 3	2
2. 4-8	5
3. 9-12	10
4. 13-18	15
5. more than 18 a week	20

When your engine runs cold, which it does on a trip of less than 2 miles, pollutant emissions are much higher. These short-distance trips produce up to 40% of vehicle pollution in urban areas.

5.6

The fuel I use in my car is _____.

1. electric car	1
2. bio-fuel (methanol, ethanol, natural gas)	2
3. a mixture of bio-fuel and unleaded gasoline	3
4. unleaded gasoline-low-octane	4
5. unleaded gasoline-high octane	6
6. leaded gasoline	10

High-octane gasoline contains compounds that are among the worst environmental pollutants: benzene, xylene, and toluene. Experts state that octane is only beneficial to certain high compression engines and is much overused.

Converting to an electric-powered vehicle can do more for the environment than just about any other individual act.

Electric cars

- emit one-tenth of the pollution of conventional cars. (The pollution comes from generating electricity at the power plant.)
- can move at highway speeds.
- have less than one-tenth the moving parts of a conventional vehicle, thus rarely break down or need maintenance.
- can easily be charged from any household outlet, including solar-generated power.

The main drawbacks to electric vehicles are that they need charging every 60 to 120 miles and they take from 2 to 8 hours to recharge. However, 95% of all vehicle trips in the United States are less 30 miles. To solve the problem of long distances, universal battery packs could

IMPACT ACTION

be developed that could be exchanged at "battery stations", something like gas stations. When your battery runs low, you would find a station and exchange it for a freshly charged battery pack, paying a fee with a credit card. Each battery could have an odometer on it to track depreciation.

> **Resources**
>
> Phillip Terpstra, *Worldwide Electric Vehicle Directory*, 1993. Tucson, AZ: Spirit Publications. 408-294-5848. $14.
>
> Electro Automotive. 1-800-537-2882. Information about converting cars from gasoline to electric.

5.7

I drive alone _____ % of my total driving miles.

1. 0-20	2
2. 20-40	4
3. 40-60	6
4. 60-80	8
5. 80-100	10

5.8

I have ____ other motor-driven vehicles (boat, snowmobile,motorcycle, seated lawn mower, airplane, recreational vehicle, and so on.)

1. 0	0
2. 1	4
3. 2	8
4. 3	12
5. 4 or more	16

5.9

I carpool for _____ % of my commuting miles annually.

1. 0	0
2. 1-20	5
3. 20-40	10
4. 40-60	15
5. 60-80	25
5. 8-100	30

IMPACT | ACTION

Resources

For van-pooling information about your area, call 1-800-223- 8774. American Council on Transportation (ACT), 808 17th St. NW, Ste.200, Washington, DC 20006. For ride-sharing information, send a self addressed, stamped envelope. You will receive a list of ride-sharing contacts for your area.

5.10

I use mass transit for _____ of my commuting miles.

1. 0	0
2 1-20	5
3. 20-40	10
4. 40-60	15
5. 60-80	25
6. 80-100	30

Van pooling and mass transit provide great opportunities for catching up on reading and work that you couldn't do while driving.

The following table illustrates the amount of pollution emitted from urban U.S. transport modes for typical work commutes:

Mode	Grams of carbon per passenger mile
rapid rail	0.3
light rail	0.4
transit bus	20.0
vanpool	36.0
carpool	70.0
automobile (single occupant)	209.0

5.11

I do ____ % of my commuting miles by bicycle or by foot.

1. 0	0
2. 1-20	5
3. 20-40	10
4. 40-60	15
5. 60-80	25
6. 80-100	30

How close you live to where you work is an important factor in your environmental impact. City dwellers have scored very well in the

IMPACT | ACTION

EnviroAccount computer audit because of this factor. Many city dwellers don't drive their cars for a week at a time, others find that they can live well without owning a car at all.

On the other hand, country dwellers often work in cities and commute long distances despite the fact that they originally moved to the countryside partly to become environmentally "sensible".

5.12

I travel ____ % of my around-town errand miles by bicycle or by foot.

1. 0	0
2. 1-20	5
3. 20-40	10
4. 40-60	15
5. 60-80	25
6. 80-100	30

Does your city have a bicycle program? Citizen pressure is important in developing programs for bicycle paths and alternative transport incentives.

Resources

Bicycle Federation of America. 202-332-6986.

5.13

For this section, add up the approximate number of miles you have flown in the last 5 years and then divide by 5 to find your 5-year average. Do not count miles flown as part of your job if you have no other choice but to fly.

I travel _____ miles per year by air.

1. 0-1,000	1
2. 1,000-5,000	2
3. 5,000-10,000	4
4. 10,000-20,000	8
5. 20,000-30,000	16
6. 30,000-50,000	32
7. more than 50,000	48

If you travel as a passenger on an airliner at near-full passenger capacity, you use about the same amount of fuel as if you drive a car (by yourself) that gets 27 miles per gallon. It is much easier to log several thousand miles per week of flying, than of driving, however.

IMPACT ACTION

5.14

When I travel by air I make _____ effort to fly by the most fuel-efficient airliner available.

1. no	0
2. a little	5
3. moderate	10
4. a good	15
5. extensive	25

New airliners are the most fuel efficient, up to 40% more so than older airliners. Some airlines will be introducing airliners with engines that have significantly lower emissions of NOx's, a pollutant that causes destruction of the ozone layer. Public inquiry will make a difference in whether airline companies decide to invest in these technologies.

Scores

Total the points for this section:

IMPACT ACTION

Section 6

Consumerism: Durable Goods

Along with driving, the buying of consumer goods is one of our major impacts on the environment, and probably the most complex and difficult to assess. The United States produces more than 20 tons of solid waste per person per year from mining, manufacturing, and farming. The industrial sector, much of which manufactures consumer goods, uses more energy than any other sector of the U.S. economy. The manufacture of some types of goods will have hundreds of times the environmental impact of that of other items in the same price range. This imbalance exists because our economic system has not developed to the point where the consumer pays for all of the environmental impacts of producing and disposing of an item. These "hidden" environmental costs, such as the disposal of dioxins generated in the production of paper, or erosion caused by the clear-cutting of forests, are called "externalities." We must work toward the day when these environmental costs are included in the cost of goods so that those who consume impact-intensive goods pay the proportional cost of environmental maintenance and cleanup. This section evaluates your general level of consumerism and your efforts to reduce the environmental impact of your consumption.

Section 6
Consumerism: Durable Goods

Circle an answer in each question. Record your points in the box provided.

IMPACT **ACTION**

6.1

I am a ______ shopper/buyer of durable goods (not counting second-hand goods). Durable goods include clothing, electronic goods, vehicles, toys, appliances, furniture, recreational items, tools, and so on, but do not count food. (Families divide the total expenditures by the number of adults.)

1. very light ($50/mo. avg. or less)	16
2. light ($50-$150/mo. avg.)	24
3. moderate ($150-$300/mo. avg)	32
4. moderately heavy ($300-$500/mo. avg.)	48
5. heavy ($500-$800/mo. avg.)	64
6. very heavy (more than $800/mo. avg.)	96

This question has to do with your role in the environmental impact of the production (resource extraction, processing, and transport) of goods. An identical question in the waste section gives Impact points for the disposal of these items.

If this question seems to be too general and you want a more thorough assessment of how your acquisition of durable goods affects the environment see the *EnviroAccount* software program (see back page of this book).

6.2

In the past 5 years, I have purchased ______ new or almost new motor vehicles. (Almost new means 2 years old or less. Do not count vehicles that were more than 2 years old when purchased.)

1. 0	0
2. 1	6
3. 2	12
4. 3	18
5. 4	24

IMPACT | ACTION

The manufacture of an automobile consumes enormous amounts of energy and raw materials, including the following:

- steel (1,500-3,500 lb.)
- plastic and composite synthetic materials (200-1,000 lb.)
- glass, aluminum, and other materials (120-200 lb.)

A single car contains four to ten times as much plastic alone as all other types of plastic that you are likely to consume within a year.

In addition, the manufacture of the average 2,370-pound car generates nearly 27 tons of waste: metals, plastics, glass, and so on.

Every time a new vehicle is bought a message is sent back to the manufacturer to produce another vehicle, which in turn sends a message back even further to extract more resources (ore, petroleum, and so on). It is sometimes better, from an environmental point of view, to keep an older car running well than to buy a new car that gets just a few more miles per gallon, depending on how much you drive.

6.3

When I want to acquire something, I make ______ effort to purchase a second-hand product by going to thrift stores and checking classified ads.

1. no	0
2. little	4
3. a moderate	8
4. a good	16
5. extensive	32

Enough people exerting enough economic demand for second-hand goods places value on well made products, which in turn gives companies incentives to produce them.

Many cities have salvage centers, businesses that specialize in all kinds of second-hand items, especially for home construction.

A new business concept called the Green Card System is gaining ground. Members bring used goods to one of many Green Card drop-off sites and receive credits, which they then use to buy goods at Green Card member shops. The Green Card System promises to grow from Southern California, where it started, to a nationwide franchise.

Resources

For Green Card System information, call 1-800-289-4733.

IMPACT	ACTION

6.4

When I want to acquire something, I make _____ effort to buy goods made from low impact, recycled, or renewable materials or, from companies that have a good environmental record.

1. no	0
2. a little	3
3. a moderate	6
4. a good	12
5. extensive	24

Resources

Alice T. Marlin ed., *Shopping for a Better World: A Quick and Easy Guide to Socially Responsible Supermarket Shopping.* New York, NY: Council on Economic Priorities, 1992. $7.95. 1-800-729-4237. The standard reference for supermarket shopping. It rates hundreds of companies and thousands of products as acceptable or unacceptable from environmental and social points of view.

Buy Recycled Campaign,*Shopper's Guide to Recycled Products.* Sacramento, CA: California Against Waste Foundation, 1991, $2. 916-443-8317

Earth Tools. 1-800-825-6460. Catalog of low-impact goods.

Co-op America. 1-800-424-2667. Hand-made imports with fair trade arrangements.

6.5

I make ______ effort to inform store managers and owners that I am interested in buying products made from low-impact, recycled, or renewable materials and from companies that have a good environmental record.

1. no	0
2. a little	2
3. a moderate	4
4. a good	8
5. extensive	16

Information is increasingly available on the ways we can choose to consume in order to bring about a lower environmental impact. Corporations are extremely sensitive to consumer choice and awareness. Do not underestimate the power of your dollar (or your telephone call).

IMPACT | ACTION

Two "green certification" companies Scientific Certification Systems and Green Seal are putting their stamp of approval on products that pass certain environmental tests. Scientific Certification Systems produces a detailed label which outlines the energy and resource costs of producing the item. Green Seal gives a simple pass/fail label for a product.

Resources

Makower, *Green Consumer*. See All-Star Bibliography.

Scientific Certification Systems. 1-800-326-3228.

Green Seal. 202-331-7337.

Scores

Total the points for this section:

IMPACT | ACTION

Section 7

Consumerism: Food and Agricultural Products

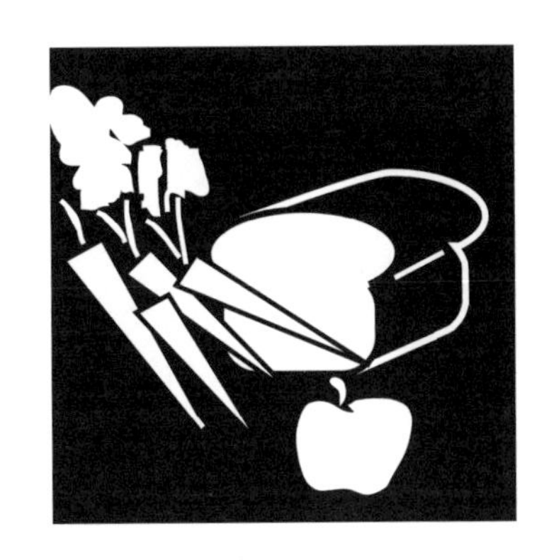

Food is our most fundamental connection to the earth. We transform the land with our agriculture and with its harvest we take in the earth's nutrients and energy to produce our own flesh. Selecting the food you eat then deserves special care. Food that is grown using organic fertilizers and ecologically based methods of pest control is worth its moderately higher price.This section evaluates your impact when buying at the grocery store, and it assesses your efforts to support ecological sustainability in food and agricultural products. Consumption of organically grown foods minimizes your Impact Points in this section, high levels of consumption of meat and extensively processed foods can combine to give you high Impact Points.

Section 7

Consumerism: Food and Agricultural Products

Circle an answer in each question. Record your points in the box provided.

IMPACT	ACTION

7.1

I ______ eat canned, frozen, and individually packaged foods.

1. never/rarely	1
2. sometimes (1 meal/wk.)	4
3. fairly frequently (2-4 meals/wk.)	8
4. frequently (5-7 meals/wk.)	12
5. usually	16

Canning, freezing, processing, and transporting foods use a lot of energy. Food packaging accounts for 30% of U.S. trash by volume.

7.2

I buy ______ of my food from bulk bins or in large packages, or mostly as unprocessed food.

1. none/very little	0
2. some	2
3. moderate amounts	4
4. a good amount	8
5. most	16

Resources

Joel Makower, *The Green Consumer Supermarket Guide: Brand-Name Products That Don't Cost the Earth*. New York, NY: Viking Penguin, 1991. $8.

IMPACT	ACTION

7.3

As a percentage of my food dollar, I buy _____ % organically grown or unsprayed food.

1. more than 40%	4
2. 30%-40%	8
3. 15%-30%	16
4. 5%-15%	24
5. 0%-5%	32

Food growing has a high environmental impact because of the large amounts of pesticides and chemical fertilizers that are applied to the land (1.5 billion pounds per year). Of these pesticides, 40% are applied so that the product will look better. By using existing organic farming techniques we could grow much of our food, perhaps most of it, without pesticides or synthetic fertilizers. Organic farming emphasizes ecological sustainability, biological diversity, and the health of soil, plants, animals, and, therefore, humans.

Ask your local grocer about organically grown produce, and check at health food stores and farmers' markets.

You can minimize your environmental impact by eating foods that are (1) low on the food chain (vegetarian), (2) minimally processed, (3) organically grown, and/or (4) locally produced.

Resources

Americans for Safe Food. 202-332-9110. For sources of organic foods, extension 348. Remember to try first to buy locally produced foods, however..

The Community Alliance with Family Farmers, *Consumer's Organic Mail-Order Directory*. Davis, CA: 1992. 1-800-852-3832. $9.95.

Sass, *Recipes from an Ecological Kitchen*. See All-Star Bibliography.

Martin Teitel, *Rainforest in Your Kitchen*. Washington, DC: Island Press, 1992. $10.95.

AgAccess Bookstore. 916-756-7177. Books on all aspects of organic farming and sustainable agriculture.

IMPACT | ACTION

7.4

As a percentage of my food dollar, I spend _____ % at the local farmers market or outlet.

Answer	Score
1. 0%-5%	0
2. 5%-15%	2
3. 15%-30%	4
4. 30%-40%	8
5. more than 40%	16

Support your local food growers. Local, small-scale farming increases ecological diversity and community self-reliance as well as reduces the need for transporting and refrigerating large amounts of fresh food grown far away.

7.5

I eat _____ pound(s) of beef per week.

Answer	Score
1. 0	0
2. less than 1/4	4
3. 1/4-1/2	8
4. 1/2-1	16
5. 1-2	24
6. 2-3	32
7. 3-4	48
8. more than 4	64

The average person in the United States eats 3 to 4 pounds of beef per week. To produce one steak takes an incredible 2,000 gallons of water.

Beef production has the highest environmental impact of any common food, particularly in the western U.S., where water is scarce and livestock graze on fragile semiarid and subalpine lands. According to Jeremy Rifkin who started the Beyond Beef Campaign, "There is no other force on earth more destructive than the cow, except for the automobile." The goal of the campaign is to influence consumers around the world to cut beef consumption by 50%. See Resources.

For those who desire beef, two beef companies specialize in beef production using sustainable grassland grazing (as opposed to grain feeding) and without using hormones and chemicals. See Resources.

IMPACT | ACTION

Resources

Beyond Beef. 202-775-1132. $25 membership.

Jeremy Rifkin, *Beyond Beef: The Rise and Fall of the Cattle Culture*. New York, NY: Dutton, 1992.

Dakota Lean Meat Company. 1-800-727-5326. Sustainably produced beef.

Coleman Beef Company. 1-800-442-8666. Sustainably produced beef.

Additional resources for meats are listed in the next question.

7.6

I eat _____ pound(s) of nonbeef meat per week (chicken, fish, pork, mutton, and so on).

1. 0	0
2. less than 1/4	2
3. 1/4 to 1/2	4
4. 1/2 to 1	8
5. more than 1	16

These foods have a lower environmental impact than beef but a much higher impact than vegetarian foods. The environmental impact of sea fishing varies. Some types of "fish harvesting" are sustainable, while many others are not, depending on the species and the region.

Many stores, especially food co-ops, sell "free range" and hormone-free chicken, poultry raised outside of cages and without the use of drugs or hormones.

Resources

Food Animal Concerns Trust (FACT). 312-525-4952. Information about where to buy meat from humanely and sustainably raised animals.

Americans for Safe Food. 202-332-9110. Information about chemicals in meat.

National Marine Fisheries Services, *Our Living Oceans: Report of the Status of* U.S. *Living Marine Resources*. Washington, D.C.: National Oceanic and Atmospheric Administration, 1992. Free. 301-713-2363. Summarizes the environmental status of each species of commercial sea fish.

IMPACT ACTION

7.7

I make _____ effort to buy textile products from organically grown or unsprayed cotton or less environmentally costly sources such as wool, linen, hemp, or recycled material, such as plastic.

1. no/very little	0
2. some	5
3. moderate	10
4. extensive	20

Unsprayed and sustainably produced, ("green") cotton is now becoming available. More pesticides are applied to cotton than to any other crop. Of all pesticides applied to crops in the United States, 50% are applied to cotton,
and these have an enormous environmental impact. The cotton-growing areas of the world, whether the lowlands of Guatemala or the San Joaquin Valley of California, are severely contaminated with pesticides and in some cases are virtual "deserts" because their ecological systems have been so disrupted.

The idea that cotton is a "natural" and "earth-friendly" product is a myth, then. In fact, while it would seem that cotton would be more environmentally benign than synthetics, this is not definitively so. For a more definitive answer we must wait for detailed studies on this subject.

Resources

Seventh Generation Company. 1-800-456-1177. Many "green" cotton products available.

Eco-Sport. 1-800-486-4326. Wholesalers of a full line of organically grown cotton products; they can tell you where you can buy their products.

Patagonia Company. 1-800-336-9090. They are making a pioneering effort to introduce clothing from sources that have a minimal environmental impact. They have just introduced a line of jackets made from recycled plastic bottles.

Scores

Total the points for this section:

IMPACT ACTION

Section 8

Consumerism: Paper and Forest Products

Forest and watershed management in western North America has become increasingly destructive and environmentally unsustainable. A 1988 Wilderness Society report *The Wasting of the Forests*, states, "Our national forest system is being destroyed by commercial logging–and taxpayers are subsidizing this disaster to the tune of almost half a billion dollars every year." (p.7; 202-833-2300). Sustainable forestry practices, which have been researched and developed, need to be implemented on a much wider scale. However, only through consumer demand for recycled paper and sustainably produced lumber products as well as citizen pressure on the U.S. Forest Service and timber corporations will this type of sustainable logging be adopted widely. This section evaluates your use of paper and forest products as well as your efforts to favor products that are less impactful.

Section 8
Consumerism: Paper and Forest Products

Circle an answer in each question. Record your points in the box provided.

IMPACT | ACTION

8.1
In my house, we subscribe to ____ daily newspapers per reader.

1. no	0
2. less than 1/4 (4 or more readers/paper)	3
3. 1/3 to 1/2 (2-3 readers/paper)	6
4. 1	9
5. 2	12
6. 3 or more	15

Paper has three major environmental impacts: (1) destruction of forests, (2) dioxin production in the chlorine bleaching process, and, (3) disposal (some 40% of landfills are paper waste).

Can you share a newspaper with a neighbor?

Resources

Elliot Norse, *Ancient Forests of the Pacific Northwest*. Covelo, CA: Island Press, 1990. 707-983-6432. $19.95. A thorough discussion of the environmental issues around forests and the timber industry.

8.2
In my house, we have _____ magazine subscriptions per reader.

1. 0	0
2. 1-2	3
3. 3-4	6
4. 4-6	9
5. 6-8	12
6. more than 8	15

Are the magazines you get crammed full of glossy advertisements for nonessential consumer goods like liquor, new cars, and perfume? Many fine magazines do not take advertising or take only minimal

IMPACT | ACTION

advertising from selected sources, which saves paper. Many of these magazines use a nonglossy format which utilizes fewer chemicals and dyes. Additionally, the editorial content of noncommercial magazines is generally more on the cutting edge. A good library or bookstore will have many of these magazines.

Resources

Utne Reader. 1-800-736-UTNE. A quarterly magazine; prints articles from and lists many cutting-edge, noncommercial journals. Available in magazine stores.

8.3

I make _____ effort to obtain my reading materials (newspapers, magazines, books, and so on) as a multiple user, that is, at libraries, in cafes, from friends, and so on.

1. no/very little	0
2. some	2
3. moderate	4
4. good	6
5. extensive	8

8.4

I am a _____ consumer of paper, other than newspapers and magazines, such as books, photocopies and so on. (Do not count paper used at work if your work gives you no other choice but to use paper.)

1. very light	3
2. light (1/8" stacked, or 1/4 lb./wk.)	6
3. moderate (1/4", or 1/2 lb./wk.)	9
4. moderately heavy (1/2", or 1 lb./wk.)	12
5. heavy (1", or 2 lbs./wk.)	15
6. very heavy	20

When you photocopy can you use a machine that does double-sided copying? Do you request recycled paper? For large numbers of such printed materials as flyers, can you reduce the size to fit four on one page and then get the paper cut into quarters? Most copy shops will make such cuts for a small fee.

IMPACT	ACTION

8.5

I make ______ effort to reduce my junk mail.

1. no	0
2. a little	1
3. moderate	2
4. good	3
5. extensive	4

Your receipt of junk mail is considered in your score even though you do not "buy" this material. Each year approximately 60 million trees are cut down for junk mail. Paper for catalogs alone caused the destruction of 74,000 acres of forest last year. Consideration should be given, however, to the fact that shopping by catalog can reduce the amount of driving you do, as well as the amount of impulse buying.

Resources

Direct Market Association. 6 East Forty-third St., New York, NY 10017. You can write and ask to be taken off mailing lists. You can reduce your junk mail by up to 75% by doing this.

Stop Junk Mail Association. 1-800-827-5549. Membership $17.50. Anti-junk mail kits available.

8.6

I make _____ effort to buy my paper products from sources that use recycled paper and to let companies know that I want them to print on recycled and chlorine free paper.

1. no	0
2. a little	1
3. moderate	2
4. good	3
5. extensive	4

To be truly recycled, paper should contain at least 50% "postconsumer" waste. Much of the paper now sold as "recycled" does not meet this criterion and may in fact have no postconsumer paper in it whatsoever. Postconsumer recycling not only saves trees but reduces the energy used in paper making by 50% and reduces air pollution by 74% and water pollution by 35%. It also reduces the load on landfills, of which paper is the largest single component.

The chlorine bleaching of pulp to make white paper is a major cause of dioxin pollution because wastewater from the bleaching process is dumped into the oceans. Last year, *Time* magazine nearly switched to chlorine-free paper, a change that would have had enormous environmental benefits because of the volume of paper used by *Time*. For

IMPACT | ACTION

various reasons, it did not make the switch. A little consumer pressure could go a long way in making this kind of change.

Resources

Recycled Products Information Clearinghouse. 703-941-4452. Information on sources of recycled and unbleached paper.

Office supply stores can order recycled paper. At least 50% post-consumer waste, nonchlorine paper is recommended.

Earthcare Paper Company. 1-800-277-2900. A full line of recycled unbleached paper available by mail order.

Makower, *Green Consumer*. See All-Star Bibliography. Household paper (products such as paper towels and toilet paper) also come recycled. Grocery stores now stock many such brands. Twelve of these companies, listed in *Green Consumer*, have received Scientific Certification System's Green Cross.

8.7

In the past 5 years, my use of forest-grown lumber products has been, on average, _____.

1. very small (100 board ft. or less)	2
2. small (100-500 board ft.)	4
3. medium (500-1,000 board ft	8
4. substantial (1,000-5,000 board ft.)	16
5. very substantial (more than 5,000 board ft.)	32

A small addition to a house might require 500 to 1,000 board feet of lumber, while a major addition might require 1,000 to 5,000 board feet. A new house generally requires 10,000 board feet.

Wood-certification programs in which timber operations are inspected as to the sustainability of their practices and given certification if they pass, are a new and important development in the United States. If you are going to have any construction done, it is important that you discuss with your contractor the options for getting sustainably produced wood.

IMPACT | ACTION

Resources

The following sources have information on sustainably produced wood and wood salvaged from buildings.

Forest Stewardship Council. Vermont. 802-434-3101.

Home Depot Stores now carry certified sustainably produced lumber. 1-800-553-3199

Global Resources Consultants. Washington, D.C.. 703-330-3889.

Institute for Sustainable Forestry. California. 707-923-4719.

Rainforest Alliance Smart Wood Program. Washington, D.C.. 212-677-1900.

Scientific Certification Systems Wood Program. California. 1-800-326-3228

Western Wood Products Association. Oregon. 503-224-3930

Major metropolitan areas all have businesses that salvage and sell wood from old buildings. Some of this wood is of a quality not available anymore.

8.8

When I have construction done, I make _____ effort to look for a contractor that specializes in environmentally sound building or to have the contractor use materials that are low impact (such as earthen materials), sustainably produced, indigenous (locally derived), or salvaged.

1. no	0
2. a little	2
3. moderate	4
4. good	6
5. extensive	8

Resources

Victoria Schomer ed., *Interior Concerns Resource Guide*. Mill Valley, CA: 1992. Interior Concerns Publishing. 415-389-8049. $40

Andrew St. John ed., *Sourcebook for Sustainable Design: A Guide to Environmentally Responsible Building Materials and Processes*. Boston, MA: Boston Society of Architects. 617-951-1433 ext. 221. $25.

Walter Spurling, *Guide to Resource Efficient Building Elements*.

Missoula, MT: Center for Resourceful Building Technology, 1993. 406-549-7678. $25. A listing of manufacturers of resource efficient building materials.

IMPACT | ACTION

8.9

I make ______ effort to avoid buying furniture and other items made from tropical wood, unless they clearly state that they are produced by sustainable methods.

1. no	0
2. a little	3
3. moderate	6
4. good	9
5. extensive	12

Tropical rainforests, which are responsible for 70% of the biomass that keeps the earth from warming (the greenhouse effect) and which are the potential source of thousands of medically important products, are being cut at a rate of 4% per year, an area the size of the state of Pennsylvania. These rainforests will be virtually gone in 20 years if we don't stop the destruction. Much of this cutting is done to meet the demand for wood products and beef in the U.S. and other developed countries (land is cleared for cattle grazing).

Common tropical woods that are endangered include Amazaque, andiroba, apitong, ebony, gaboon, iroko, jelutong, kapur, kempas, lauan/meranti, lignumvitae, macawood, mahogany, Merbau, Nyatoh, Mexican oak, obeche, piquia, ramin, sapele, utile, and wenge.

North American hardwoods that are decorative and durable and can substitute for tropical woods include ash, red birch, cherry, bird's-eye maple, myrtle, red and white oaks, English brown oak, pearwood, and walnut.

Resources

Rainforest Alliance Smart Wood Program. Washington, D.C.. 212-677-1900.

Janet Marinelli with Robert Kourik, *The Naturally Elegant Home.* Boston: Little, Brown, 1992. 1-800-343-9204. $45.

See also Resources, 8.5.

IMPACT ACTION

8.10

When I acquire an undomesticated pet such as a bird or a snake, I make _____ effort to make sure that it was bred in captivity and not taken from the wild.

1. no	4
2. a little	3
3. moderate	2
4. good	1
5. I do not have any undomesticated pets.	0

This question can also apply to house plants, such as cacti and to some specialty foods, such as hearts of palm, turtle eggs, and so on.

Resources

National Wildlife Federation. 202-797-6800.

Scores

Total the points for this section:

IMPACT ACTION

Section 9

Toxics

The average U.S. home generates 15 pounds of hazardous waste each year. The pie chart shows the breakdown of types of waste.

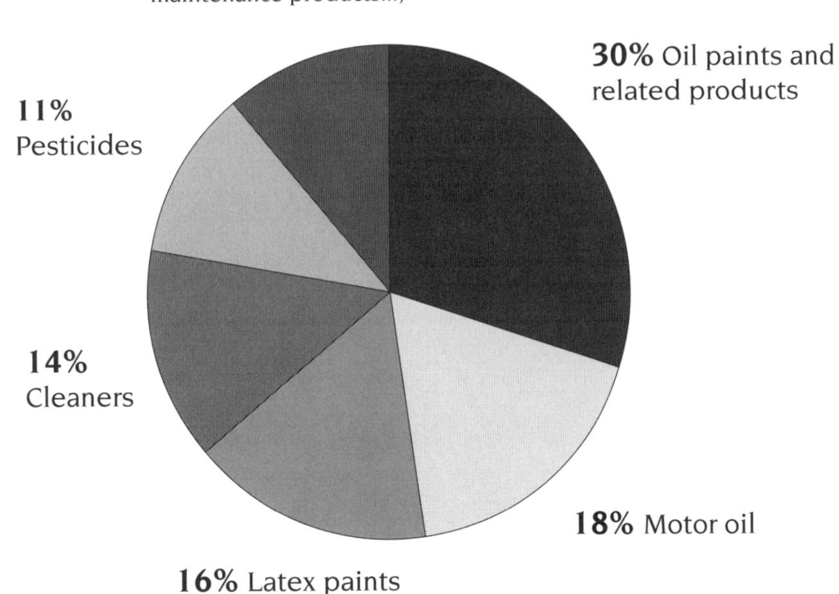

Hazardous materials, when improperly disposed of in landfills, can leach into waterways and seriously damage ecosystems and endanger human health. This section evaluates your efforts to minimize the use of toxic products and to dispose of them properly.

Section 9
Toxics

Circle an answer in each question. Record your points in the box provided.

IMPACT	ACTION

9.1

I make ______ effort to keep hazardous wastes (old batteries, solvents, used motor oil, paints, pesticides and so on) from going into the garbage or sewer and to take these items to a hazardous waste collection point.

1. no	32
2. a little	24
3. moderate	16
4. good	8
5. extensive	4

The Environmental Protection Agency has designated the following chemicals as toxic enemies 1 through 16. Check labels for these products. Many, however, are "hidden" secondary products in the manufacture of products you buy, five of the sixteen for example, are hidden in the manufacture of plastics.

1. benzene
2. cadmium and its compounds
3. carbon tetrachloride
4. chloroform
5. chromium and its compounds
6. cyanides
7. dichloromethane
8. lead and its compounds
9. nickel and its compounds
10. methyl ethyl ketone
11. tetrachloroethylene
12. toluene
13. 1,1,1-trichloroethane
14. trichloroethylene
15. mercury and its compounds
16. xylene(s)

Nickel/cadmium (NiCad) batteries thrown away by consumers are the single largest source of cadmium pollution (toxic enemy no. 2) in the environment. Most rechargeable batteries are NiCad, including built-in batteries such as those in camcorders and other rechargeable electronic devices. Never throw away NiCad batteries or devices that contain rechargeable batteries. Take them to a hazardous waste collection site for proper disposal.

IMPACT | ACTION

Resources

The National Response Center Hotline. 1-800-424-8802. Call to report a toxic spill.

Real Goods Trading Company. 1-800-762-7325. Non-NiCad rechargeable batteries (nickel metal hydride) available.

9.2

My lawn, garden, or home is _____ sprayed with herbicides or pesticides.

1. never	0
2. rarely	2
3. occasionally (once/yr.)	4
4. regularly (once/mo.)	6
5. frequently (once/wk.)	8

Resources

National Coalition against the Misuse of Pesticides. 202-543-5450
Rachel Carson Council. 301-652-1877.

Rhonda M. Hart, *Bugs, Slugs, and Other Thugs: Controlling Garden Pests Organically*. Pownal, VT: Storey Communications, 1991.1-800-441-5700. $9.95.

Stuart Franklin, *Building a Healthy Lawn: A Safe and Natural Approach*. Pownal, VT: Storey Communications, 1988. 1-800-441-5700. $9.95.

William Olkowski et.al., *Common Sense Pest Control*. Newtown, CT: Taunton Press, 1991. $40.

Necessary Trading Company. 703-864-5103. Organic pest control products available.

9.3

I make _____ effort to tolerate occasional insects or weeds around my home and to control them using environmentally friendly methods.

1. no	0
2. a little	1
3. moderate	2
4. good	4
5. extensive	8

IMPACT | ACTION

Resources

AgAccess Bookstore. 916-756-7177. Books about environmentally friendly pest control.

9.4

I am a _____ user of strong household maintenance products such as cleansers, polishes, and so on.

1. negligible	1
2. small	3
3. moderate	6
4. frequent	9
5. heavy	12

The harmful ingredients here are chlorine, phenols, formaldehyde, cresol, and the benzenes, all commonly found in household maintenance products. Look for these on the label (they may be hidden in long words like p-dichlorobenzene or pentachlorophenol). Many of these cause cancer and can find their way back into your drinking water if disposed of in the sewer system. See Section 9.5 for alternative products.

Resources

Environmental Hazards Management Institute. P.O. Box 932, Durham,NH 03824. $6 for an information packet "Household Hazardous Waste Wheel."

See also Resources, 9.6

9.5

I make _____ effort to use environmentally friendly, low-toxicity cleaning and household maintenance products or methods.

1. no	0
2. a little	4
3. moderate	8
4. good	16
5. extensive	32

IMPACT | ACTION

Resources

Scientific Certification Systems. 1-800-326-3228. Has given its Green Cross certification to some of the many "green" cleansers now on the market. Look for the Green Cross on the product, or call for information.

Makower, *Green Consumer*. See All-Star Bibliography. Lists Green Cross certified cleansers. Page 166 gives information about the old standbys for environmentally safe cleaning, including Arm & Hammer baking soda, Bon Ami polishing cleanser, Dr. Bronner's pure castile soap, Fels Naptha soap, Ivory soap, and Twenty Mule Team Borax. Consult page 166 of The Green Consumer for more information.

Anne Berthold-Bond, *Clean & Green: The Complete Guide to Nontoxic and Environmentally Safe Housekeeping*. Woodstock, NY: Ceres Press, 1990. $10.95.

9.6

I am a _____ user of solvent-based compounds such as paints, paint removers, varnishes, cleaning chemicals, and so on.

1. negligible	1
2. small	4
3. moderate	8
4. moderately heavy	16
5. heavy	24

Resources

Household Hazardous Waste Project, *Guide to Hazardous Products around the Home*, 2d ed. Springfield, MO: Household Hazardous Waste Project, 1989. $9.95. 417-889-5000.

U.S. Environmental Protection Agency Public Center. 202-382-2080.Or call one of ten regional EPA offices, listed in your phone book.

See also Resources, 9.4

9.7

I make _____ effort to use non-solvent-based paints and other compounds for my home projects.

1. no	0
2. a little	2
3. moderate	4
4. good	8
5. extensive	16

IMPACT | ACTION

Resources

These sources have environmentally and health friendly paints, varnishes, and so on.

AFM Enterprises. 714-781-6860.

Auro Natural Paints. 916-753-3104 (West Coast), 617-489-6747 (East Coast).

Baubiologie Hardware. 602-445-8225.

Livos Plantchemistry. 505-438-3448

9.8

I am a ______ user of dry cleaners.

1. negligible 0
2. small 1
3. moderate 2
4. frequent 3
5. heavy 4

Dry cleaning generates large amounts of toxic waste from the use of perchlorethylene, or "perc", the main cleaning agent. New methods of dry cleaning are being developed using biodegradable cleaners. Ask your dry cleaner about the process, or call the Environmental Protection Agency (see Section 9.7, Resources).

9.9

I use nonphosphate laundry soap.

1. always 1
2. most of the time 2
3. half of the time 3
4. sometimes 4
5. never/rarely 5

Phosphate is not a toxic compound, but it is a pollutant because it stimulates algae growth which deoxygenates waterways, leading to suffocation of fish. Many brand-name detergents now available in supermarkets have no phosphate in them. Check the ingredient label.

IMPACT | ACTION

9.10

I have air-conditioning _____.

1. in neither my car nor my house	0
2. in my car or house, but it's non-CFC	2
3. in my car or house	4
4. in more than one car	6
5. in my car(s) and my house	8

Chlorofluorocarbons (CFCs) which are used in air-conditioning systems, are a major cause of the loss of the ozone layer. With 5% of the world's population, the United States contributes 29% of the world's CFCs to the atmosphere.

Most damaging to the ozone layer are the R-12 CFCs. Until recently air conditioning and refrigeration systems contained R-12 CFCs. New systems use non-CFCs. You can arrange to have CFCs from your car recycled, and replaced with non-CFC compounds. Most urban areas have car air-conditioner repair businesses that can drain and then recycle or destroy the dangerous compounds and replace them with more environmentally friendly ones.

Resources

Environmental Defense Fund. 212-505-2100. Call for information about recycling and replacing CFCs and reducing your use of ozone-destroying chemicals.

9.11

When I want to purchase an item such as foam-filled furniture, I make ______ effort to find out if the product has been manufactured without using CFCs.

1. no	0
2. a little	4
3. moderate	6
4. good	10
5. extensive	16

Foams that are manufactured without CFCs are Ultracel, Hyperlite, and Geolite.

Fire extinguishers that use halons have CFCs. Dry chemical or sodium bicarbonate extinguishers do not use halons and are commonly available.

IMPACT | ACTION

9.12

I make sure that my motor oil gets recycled, that my old car batteries get recycled, and that my drained antifreeze is either recycled or taken to a hazardous waste collection point.

1. all of the above are true	1
2. two of the above are true	2
3. one of the above are true	3
4. none of the above are true	6

Used motor oil is considered hazardous waste and has a high disposal cost.

According to the Environmental Protection Agency the largest single source of oil pollution fouling our nation's waters is not tanker spills (including the *Exxon Valdez*), but used motor oil dumped by home mechanics. Additionally, each day Americans throw away 90,000 oil filters.

Resources

System One Filtration. 209-687-1955. High-quality, permanent oil filters available to reduce the need to change oil by 80% and eliminate the need to throw away filters.

9.13

I buy recycled (re-refined) motor oil for _____ % of my oil needs.

1. 0	0
2. 1-25	1
3. 25-50	2
4. 50-75	3
5. 75-100	4

Recycling (re-refining) motor oil has been compared to washing clothes: You don't need to throw away clothing the first time it gets dirty. Re-refined motor oil is available at many of the larger discount auto stores. Synthetic oils, which reduce the need for oil changes by increasing mileage per oil change, are also available.

IMPACT | ACTION

> **Resources**
>
> Amsol. 1-800-777-8491. Synthetic motor oils.

9.14

The sewer system my house is connected to causes _____ problems by leaking or discharging raw sewage into waterways or groundwater systems.

1. no	4
2. little or infrequent	8
3. occasional	16
4. frequent	32
5. frequent and serious	48

This question is mostly of concern for people with septic tank sewage systems.

> **Resources**
>
> U.S. Environmental Protection Agency Small Flows Clearinghouse. 1-800-624-8301. Call for information about how to improve or replace your septic system.

Scores

Total the points for this section:

IMPACT | ACTION

Section 10

Waste, Packaging, Single-Use Items, and Recycling

Americans produce more than twice as much waste (around 4.5 pounds per person per day) as our nearest "competitors" in waste production, Germany and Japan (1.7 pounds), and up to 30 times more than individuals in many other countries.

Eighty percent of our waste goes to landfills, more than half of which will be filled up within 10 years. We recycle only about 10% of our waste. This section evaluates your efforts to minimize waste and to recycle.

Section 10

Waste, Packaging, Single-Use Items, and Recycling

Circle an answer in each question. Record your points in the box provided.

	IMPACT	ACTION

10.1

I am a _____ shopper/buyer of packaged or throw-away goods. (Families: Divide total expenditures by the number of adults.)

1. very light ($50/mo. avg. or less)	5
2. light ($50-$150/mo. avg.)	10
3. moderate ($150-$300/mo. avg)	15
4. moderately heavy ($300-$500/mo. avg.)	20
5. heavy ($500-$800/mo. avg.)	25
6. very heavy (more than $800/mo. avg.)	30

Do you recognize this question? It is virtually the same as the one on durable goods (Section 6.1), since durable goods are almost always packaged. Those points were for the manufacture of the products, while these are for the throwing away of the products and their packaging. These are the two main ways that we impact the environment with our buying.

Packaging accounts for about 60% of the paper in landfills, or about 25% of total landfill space.

10.2

I make _____ effort to avoid buying overpackaged products and to let companies know that I will not buy their products if they overpackage.

1. no	0
2. a little	3
3. moderate	6
4. good	9
5. extensive	12

IMPACT	ACTION

The solid waste flow for the average person in this country is as follows:

41.0% paper and paperboard	20 lb./yr.,	10.0 lb./wk.
17.9% yard wastes	230 lb./yr.,	4.4 lb./wk.
8.7% metals	111 lb./yr.,	2.1 lb./wk.
8.2% glass	105 lb./yr.,	2.0 lb./wk.
8.1% textiles, rubber, wood, leather	103 lb./yr.,	2.0 lb./wk.
7.9% food wastes	100 lb./yr.,	1.9 lb./wk.
6.5% plastics	83 lb./yr.,	1.6 lb./wk.
1.7% miscellaneous	20 lb./yr.,	0.4 lb./wk.
Total	**1,272 lb./yr.,**	**24.4 lb./wk.**

With a reasonable program of reducing, reusing, recycling, and composting, this total can be reduced to 300 to 400 pounds.

The three Rs of waste reduction are reduce, reuse (refurbish and repair) and recycle.

10.3

I eat _____ fast-food meals per month.

1. 0-3	2
2. 4-7	4
3. 8-15	6
4. 16-30	8
5. more than 30	10

Eating in restaurants that use washable dishware (rather than disposable) is generally even more environmentally efficient than eating at home, because of volume and scale. If you do eat in fast food restaurants, you can ask to be served without cardboard and polystyrene foam containers.

10.4

When alternatives are available, I make _____ effort to avoid eating food or drinking beverages served in single-serving, throw-away containers.

1. no	0
2. a little	2
3. moderate	4
4. good	6
5. extensive	8

IMPACT **ACTION**

Have you thought about the fact that a plastic spoon or fork has an average use life of about 3 minutes and a geological life in the landfill of a century or more?

10.5

I carry utensils to use instead of throw-aways at snack bars and food gatherings _____.

1. never	0
2. mug, sometimes	2
3. mug, most of the time	4
4. mug, spoon, fork-sometimes	6
5. mug, spoon, fork-most of the time	8

Light and durable utensils for carrying around are available from camping stores.

> **Resources**
>
> Recreational Equipment Incorporated (REI). 1-800-426-4840. Carry-around utensils available.

10.6

I am a ______ consumer of canned or bottled beverages during summer.

1. light (0-2 /wk.)	1
2. moderate (3-6 /wk.)	2
3. moderately heavy (7-14 /wk.)	3
4. heavy (15-21 /wk.)	4
5. very heavy (more than 21 /wk.)	5

Following is a list, based on one developed by Makower in *The Green Consumer*, of types of packaging (or containers) from best to worst.

Best:	None
Very good:	Reusable and refillable containers
Good:	Packages made from recycled and recyclable materials
Fair:	Packages made from multiple layers of recycled and/or recyclable materials
Bad:	Unrecyclable packages made from multiple layers of material, or from composite materials

IMPACT | ACTION

Try carrying a water bottle in your car and in your day bag to avoid having to buy beverages from single-use containers.

10.7

I take my own cloth or used grocery bags to the grocery store for bagging items _____.

1. never	0
2. rarely	1
3. sometimes	2
4. most of the time	3
5. all of the time	4

Do not underestimate the effect you can have on other people when they see you taking this kind of action.

10.8

I recycle _____ % of my aluminum.

1. 80-100	2
2. 60-80	4
3. 40-60	6
4. 20-40	8
5. 0-20	10

Recycling aluminum saves 95% of the energy cost of production, and reduces 95% of the air pollution and 97% of the water pollution.

Resources

Earth Works Group, *The Recycler's Handbook*. Berkeley, CA: Earth Works Press, 1990. 415-841-5866. $4.95.

National Recycling Coalition. 202-625-6406.

For information on recycling of aluminum, call 202-862-5100.

For information on recycling steel and other metals, call 202-466-4050.

IMPACT	ACTION

10.9

I recycle _____ % of my glass.

1. 80-100	2
2. 60-80	4
3. 40-60	6
4. 20-40	8
5. 0-20	10

Recycling glass saves 50% of the raw materials, 50% of the water, and 33% of the energy, and it reduces air pollution by 20%. Using refillable bottles saves four times the energy of using new bottles and 100% on materials. In addition, states with bottle deposit laws have 30-40% less litter.

When depositing used bottles for recycling, keep colors separate or refrain from breaking bottles so that the colors can be separated at the recycling point.

10.10

I recycle _____ % of my paper.

1. 80-100	2
2. 60-80	4
3. 40-60	6
4. 20-40	8
5. 0-20	10

Major recycling classifications include newspaper, white paper, colored paper, corrugated cardboard. For every 150 pounds of paper you recycle, you save one tree.

10.11

I recycle ______ % of my recyclable plastic.

1. 80-100	2
2. 60-80	4
3. 40-60	6
4. 20-40	8
5. 0-20	10

When faced with a packaging choice, and you want to be environmentally friendly, choose aluminum or glass over plastic.

IMPACT | ACTION

Following are four reasons to reduce your plastic consumption: (1) The manufacture of plastic generates 5 of the 6 hazardous wastes that the EPA has designated as the worst problem chemicals (propylene, phenol, ethylene, polystyrene, and benzene). (2) Of the 10 worst polluting companies in the United States, 6 are involved in the manufacture of plastics or plastic-related compounds. (3) Plastic can last for centuries in landfills.(4) Most of the plastic you "recycle" does not actually get "recycled". Some of it may find its way into asphalt or into product filler or other low-value products. Only a small percentage of recycled plastic goes back into making containers. This "closes the loop", one of the goals of the movement toward an environmentally friendly industry.

Resources

For information on recycling of plastics, call the Center for Plastics Recycling Research at 908-932-4402.

For information on recycling polystyrene (Styrofoam), call the Association for Foam Packaging Recyclers at 1-800-944-8448.

10.12

I recycle ______ % of my corrugated cardboard.

1. 80-100	2
2. 60-80	4
3. 40-60	6
4. 20-40	8
5. 0-20	0

Ask your waste-removal company if it recycles the cardboard you throw out. Large stores often have bins for cardboard outside the back of the store, and this cardboard is recycled.

10.13

I compost _____.

1. neither kitchen nor yard wastes	12
2. kitchen or yard wastes, but not both	6
3. both kitchen and yard wastes	0

Find out if your city has a composting program for yard wastes or if it takes them to landfills. If it composts yard wastes, then answer either 2 or 3 above.

Composting food and other kitchen wastes can decrease the weight of your garbage by 30%-50%. Yard wastes are the second biggest compo-

IMPACT	ACTION

nent of landfills, after paper. In the fall, 75% of garbage taken to landfills is leaves. After being taken to landfills, these organic materials can never be recycled back to the land because they become mixed with toxic garbage.

Before trying backyard composting read about how to do it properly, or buy a compost bin.

Resources

Harmonious Technologies, *Backyard Composting: Your Complete Guide to Recycling Yard Clippings*, $6.95 (800-345-0906)

Deborah L. Martin and Grace Gershung, eds., *Rodale Book of Composting: Easy Methods for Every Gardener.* Emmaus, PA: Rodale Press, 1992. $21.95

Smith & Hawken. California. 1-800-623-3800. Composting bins available.

7th Generation. Vermont. 1-800-456-1177. Composting bins available.

10.14

When I have any construction done, or when I do a major house cleanup, or when I move, I make _____ effort to minimize the amount of waste going into dumpsters.

1. no	0
2. a little	2
3. moderate	4
4. good	6
5. extensive	8

Home projects such as major cleanups and renovations can generate huge amounts of waste. Dealing with environmental concerns can be difficult during these times because they are often times of stress and hurry. So plan ahead.

10.15

I buy retreaded tires for _____ % of my automotive tire needs.

1. 0	0
2. Up to 25%	2
3. 25% to 50%	4
4. 50% to 75%	6
5. 75% to 100%	8

IMPACT | ACTION

Tire disposal is a big problem, with more than 250 million tires a year being added to enormous piles. In all, an estimated 2 billion to 3 billion tires are stockpiled in the United States. By buying retreaded tires, you keep tires out of these dumps.

According to John Serumgard, chair of the Scrap Tire Management Council, retreaded tires are every bit as safe as new tires. Most tire-related accidents involve underinflated or bald tires.

Resources

For information about high-quality retreaded automobile tires with handling and durability virtually the same as new tires, call the following:

Tire Retread Information Bureau. 202-625-3247

American Retreaders Association. 1-800-426-8835.

Scores

Total the points for this section:

IMPACT | ACTION

Section 11

Environmental Advocacy

There are three main ways to bring about better environmental management policies by corporations: (1) through selective buying, (2) by advocating government regulation, and (3) through direct actions such as making telephone calls, writing letters or fostering public awareness through activities such as leafleting or picketing. This section counts what you do in (2) and (3). Do not underestimate the effectiveness of writing letters and doing public advocacy work for the environment. When a certain critical number of people get involved in an issue, change comes about.

Section 11
Environmental Advocacy

Circle an answer in each question. Record your points in the box provided.

IMPACT | **ACTION**

11.1

I make _____ effort to keep informed on state and federal legislative activities relevant to the environment and to make phone calls to my legislators at critical times.

1. no	0
2. a little	4
3. moderate	8
4. good	12
5. extensive	16

Resources

The following are hotlines that will inform you about current environmental legislation that needs citizen input.

Sierra Club. 202-547-5550.

Audubon Society. 202-547-9017.

National Wildlife Federation. 202-797-6655.

11.2

I write approximately _____ letter(s) per year to my government representatives regarding environmental issues.

1. 0	0
2. 1	10
3. 2-3	12
4. 4-6	14
5. more than 6	16

Writing just one letter per year makes a big difference. Note how this is reflected in the points.

IMPACT	ACTION

Resources

20/20 Vision. 1-800-347-2767. $20/yr. This is an excellent organization for time-short individuals. You receive an "action alert" every month about important bills in Congress on both environment and peace issues. The program is set up so that you spend only 20 minutes a month writing.

11.3

I vote _____% of the time. (This counts all elections: federal, state, local, and, if you are a student, school.)

1. 0-20%	0
2. 20%-40%	4
3. 40%-60%	8
4. 60%-80%	12
5. 80%-100%	16

Resources

League of Conservation Voters. 202-785-VOTE. Information on elected officials and candidates for state and federal offices. The League also has a book with this information: *Vote for the Earth*, $4.95.

Consumer Information Center in Pueblo, *Consumer's Resource Handbook*. Pueblo, Co. 719-948-3334. Information on writing to corporations about environmental issues.

11.4

I volunteer ______ hours per year to activities for the environment, such as cleanups, tree planting, community organizing, committees, fund-raising, and so on.

1. 0	0
2. 1-10	4
3. 10-25	8
4. 25-50	12
5. 50-100	16
6. 100-200	24
7. 200-400	32

IMPACT	ACTION

Resources

Richard Nilsen ed., *Helping Nature Heal: An Introduction to Environmental Restoration*. Berkeley, CA: Ten Speed, 1991. $16. An excellent information source for environmental restoration work.

Maritza Pick, *How to Save Your Neighborhood, City, or Town*. San Francisco, CA: Sierra Club, 1992. 415-923-5500. $12.

Rhode Island Sea Grant Information Office, University of Rhode Island Bay Campus, Narragansett, RI 02882-1197. Send $1 for a *National Directory of Citizen Volunteer Environmental Monitoring Programs*. This directory contains valuable information on citizen monitoring programs.

Global Releaf Program. 202-667-3300. Information on tree planting.

11.5

I donate $____ per year to environmental groups or environmental concerns. (Include money spent on membership dues and subscriptions for environmental organizations.)

1. 0-10	0
2. 10-25	3
3. 25-50	6
4. 50-100	9
5. 100-200	12
6. more than 200	16

Resources

Susan D. Lanier-Graham, *The Nature Directory: A Guide to Environmental Organizations*. New York: Walker Company, 1991. 1-800-ATWALKER $15.95. A directory that summarizes 130 environmental organizations.

Environmental Federation of America. 1-800-673-8111. Information on environmental organizations.

Council of Better Business Bureaus' Philanthropic Advisory Service (PAS). 703-276-0100.

National Charities Information Bureau. 212-929-6300.

IMPACT | ACTION

11.6

I make _____ effort to communicate my environmental concerns to key people such as local city council persons, store managers, administrators, business owners, and so on.

1. no	0
2. a little	4
3. moderate	8
4. good	12
5. extensive	16

11.7

Of my financial investments, _____ % are in "green" portfolios.

1. 0 (or no investments)	0
2. 1-20	2
2. 20-40	4
3. 40-60	8
4. 60-80	12
5. 80-100	16

In the rapidly growing field of "green investing", dozens of funds exist that invest only in corporations with good environmental records.

If you do not invest, you can still have an impact on investments. Virtually all city, county, and state governments have investments, as does your pension fund.

Resources

Co-op America, A *Socially Responsible Financial Planning Guide*. Washington, D.C.: Co-op America, 1992. 1-800-424-2667. $5.

Myra Alperson et.al.,*The Better World Investment Guide*. New York: Prentice Hall, 1991. $22.95.

The Social Investment Forum. 612-333-8338. An information clearinghouse. Guide and newsletter available.

The Green Money Guide, Quarterly newsletter of socially and environmentally responsible investing. $25/yr. 509-328-1741.

Scores

Total the points for this section:

IMPACT | **ACTION**

Section 12

Respect for the Land

This section counts how you have a direct impact on the land. Although our indirect impacts are generally of a higher magnitude, we also directly affect the land and its ecosystems in many ways. Try to visualize how you affect the ecosystem you live in. When you do this, small actions that seem insignificant, like getting rid of some "unsightly" shrubs in a corner of your backyard, may actually turn out to be harmful to an ecosystem. Do the shrubs provide habitat for birds? Do they provide an overwintering site for certain beneficial insects that eat garden pests during the summer? These questions may take years for you to answer, but being aware of them is a first step.

Section 12
Respect for the Land

Circle an answer in each question. Record your points in the box provided.

	IMPACT	ACTION

12.1
The soil erosion on my property is _____.

1. insignificant/N.A.	0
2. very small	1
3. small	2
4. moderate	3
5. often fairly serious	6
6. often very serious	12

Resources

Soil Conservation Service. Each county and region has an office. Check the federal government listing of your telephone book.

See also Resources, Section 11.4. Restoration work.

12.2
I do _____ miles of off road driving or motorcycling per year on land that has no existing track.

1. 0	0
2. 1-25	2
3. 25-100	4
4. 100-500	6
5. more than 500	8

IMPACT ACTION

12.3

When I hike or bike in natural areas, I make _____ effort to stay on existing trails and to make a minimum impact, especially on land that is wet or fragile.

1. no	0
2. a little	2
3. moderate	4
4. good	6
5. extensive	8

12.4

I make _____ effort to leave part of my land, whether backyard or extensive property, to grow wild for wildlife habitat.

1. no	0
2. a little	2
3. moderate	4
4. good	6
5. extensive	8

Resources

National Wildlife Federation "Backyard Wildlife Habitat Program." 703-790-4556.

Sara Stein, *Noah's Garden: Restoring the Ecology of America's Backyard*. Boston, MA: Houghton-Mifflin, 1993.

Xerces Society. 503-222-2788. Call for pamphlets and books on butterfly gardening. Or see Xerces Society, *Butterfly Gardening: Creating Summer Magic In Your Garden*. Portland, OR: Xerces Society. $18.95.

Bill Merilees, *Attracting Backyard Wildlife: A guide for nature lovers*. Stillwater, MN: Voyageur Press, 1989. $14.95.

Scores

Total the points for this section:

IMPACT ACTION

Section 13
Livelihood

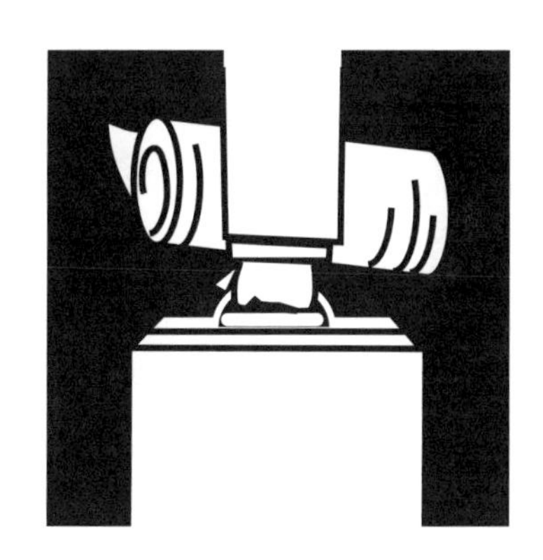

For those who are able, one of the most important ways to make a difference to the earth is to leave organizations that are on environmentally destructive treadmills. Many companies, though still a small minority, are making fundamental changes in their operations in order to minimize their environmental impact. Making changes within your work can be just as effective as changing careers. One suggestion to a supervisor could bring about an environmental impact reduction that would dwarf all the changes you make in your personal life.

This small section has two questions about livelihood: one regarding changing your work and the other regarding making changes within your work.

Section 13
Livelihood

Circle an answer in each question. Record your points in the box provided.

IMPACT	ACTION

13.1

I have made or I am making ______ effort to develop a livelihood that is earth-friendly or is with an earth-friendly company or organization.

1. no/very little	0
2. a little	8
3. moderate	16
4. good	32
5. extensive	48

Resources

The Environmental Careers Organization, *The New Complete Guide to Environmental Careers*. Washington, DC: Island Press, 1993. 1-800-828-1302. $15.95.

Susan Cohn, *Green at Work: Finding a Business Career that Works for the Environment*. Washington, DC: Island Press, 1992. 1-800-828-1302. $16

Dominguez and Robin, *Your Money or Your Life*. See All-Star Bibliography. This book has gotten rave reviews from many respected people, such as Ralph Nader. Stepping off of the workaholic/consumerholic treadmill may be the biggest step for the environment you can make. If you or someone you know is ready for this kind of change, this is a good book to read.

Utne Reader, July/August 1991. 1624 Harmon Place, Minneapolis, MN 55403. $4 prepaid. An excellent series of articles on changing to less stressful and more environmentally and socially benign jobs, careers, lifestyles, and so on.

Community Jobs. 1516 P St. NW, Washington, DC 20005. 202-667-0661. A nationwide listing of socially and environmentally responsible jobs and internships. Monthly publication.

IMPACT | ACTION

13.2

I have made or am making _____ effort to bring about changes that reduce the environmental impact of my company, department, or office.

1. no/very little	0
2. a little	4
3. moderate	8
4. good	12
5. extensive	16

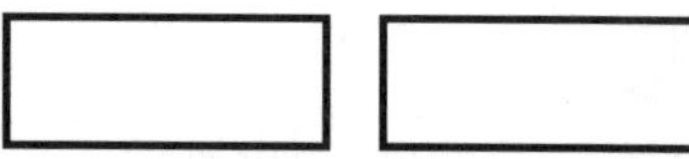

Scores

Total the points for this section:

IMPACT | ACTION

Section 14

Family Planning

There are only two questions in this last section, but they are enormously important ones; note the magnitude of the points. The number of children you have is your biggest environmental impact. This is not to say that having children is bad. In fact, one or two children brought up by anyone who cares enough to use this book is most probably a positive contribution to the world. We need educated and aware people to lead the world and make their mark on the next generation. Zero population growth can be achieved with two children per couple.

Section 14
Family Planning

Circle an answer in each question. Record your points in the box provided.

IMPACT | ACTION

14.1

I plan to have ____ children. (If you already have children, ask them to answer this question.)

1. 0	0
2. 1	10
3. 2	50
4. 3	500
5. 4 or more	5,000

Population growth is exponential, which means that very large numbers are reached after just a few generations. Here is an example of exponential population growth: My cat has 7 kittens. Each of them has 7 kittens, and in 2 years there are 49 kittens. In 7 years, there are .5 million kittens.

Limiting population growth is especially important in the United States. **One American** uses as much energy as: 2 Germans, 3 Swiss, 6 Yugoslavs, 9 Mexicans, 16 Chinese, 19 Malaysians, 53 Indians, 109 Sri Lankans, 438 Malians, or 1072 Nepalese. This illustrates why it is important that we in the United States focus on our own population problem before worrying about other countries.

Resources

Zero Population Growth. 202-332-2200.

IMPACT ACTION

14.2

I make _____ effort to raise my children in an environmentally sensible way and to set a good example for them.

1. very little 0
2. some 6
3. moderate (sporadic) 12
4. moderate (sustained) 24
5. extensive (sustained) 48

Scores

Total the points for this section

IMPACT ACTION

All-Star Bibliography

I recommend these six books, chosen from among hundreds, to help you make changes in your *EarthScore*. Between them, they cover most of the 14 areas of *EarthScore*.

Makower, Joel. *The Green Consumer*. New York: Penguin, 1993. $11. The best all-around book on how to go about being an environmentally conscious consumer. It has detailed information about many categories of products, from baby products to furniture.

Sardinsky, Robert, *The Efficient House Sourcebook*. Snowmass, CO: Rocky Mountain Institute, 1992. 303-927-3851. $15. Lists many hundreds of sources of information and products for the resource efficient home. RMI is the well-respected institute founded by energy guru Amory Lovins.

Sass, Lorna J. *Recipes From an Ecological Kitchen*. New York: William Morrow, 1992. $25. A complete vegetarian cookbook, with explanations of how to become informed about, buy, store, and prepare ecologically sound and delicious foods. Food is the most fundamental way we relate to the environment.

Dominguez, Joe, and Vicki Robin, *Your Money or Your Life: Transforming Your Relationship With Money and Acheiving Financial Independence*. New York: Penguin, 1992. $20. This book has received rave reviews from respected people like Ralph Nader and Dr. Bernie Siegel. It is about achieving financial independence as well as resource efficiency.

Wilson, Edward O. *The Diversity of Life*. Cambridge, MA: Harvard University Press, 1992. $32.95. The dean of ecologists gives us a thorough and very readable introduction to ecology and biodiversity. This book explains why it is urgent that we care for and restore the remaining natural systems of the earth.

Stewart, Mary Lou, *Stewarts Environmental Directory* 1993. Los Angeles, CA: Mary Lou Stewart, 1993. $12.95. Lists over 12,000 environmentally involved companies and organizations with over 850 categorized listings such as Environmental Architects, Biological Pest Control, Organic Fabrics, Lumber, Environmental Investments, and so on. 604-948-1566.

	IMPACT	ACTION

EarthScore Totals

Total section scores:
Total the points for all the sections:

Section	IMPACT	ACTION
Household Energy: General	______	______
Household Energy: Winter	______	______
Household Energy: Summer	______	______
Water	______	______
Transportation	______	______
Consumerism: Durable Goods	______	______
Consumerism: Food & Agricultural Products	______	______
Consumerism: Paper & Forest Products	______	______
Toxics	______	______
Waste, Packaging, Single-Use Items, and Recycling	______	______
Environmental Advocacy	______	______
Respect for the Land	______	______
Livelihood	______	______
Family Planning	______	______

Your EarthScore Total
Now with the total Impact and Action Points for all sections you can fill out the EarthScore Chart.

IMPACT	ACTION

EarthScore Chart

Find out which of the following 8 categories you fall into:

Impact Points

Rating Points	Points
Eco-Titan	150 or less
Eco-Hero	151-225
Eco-Mentor	226-275
Eco-Average Citizen	276-350
Eco-Slowpoke	351-450
Eco-Frankenstein	451-600
Eco-Sherman Tank	601-750
Eco-Tyrannosaurus rex (bound for extinction)	over 750

Action Points

Rating Points	Points
I could do better!	0-100
Good	101-200
Very Good	201-300
Excellent	301-400
Awesome (Earth Action Hero)	more than 400

Congratulations on finishing *EarthScore*! It is very important that you save your *EarthScore* book and complete *EarthScore* again 6 months to 1 year from now. You can use a different colored pencil the second time.

EnviroAccount Software

For the first time, virtually all of your impacts on and actions for the environment can be quantified and made tangible with this new, easy to use personal computer software program. *EnviroAccount* does a full accounting of your links to the environment, plus it gives you the latest information available on how to reduce your environmental impact.

- *EnviroAccount* prints out your own personal environmmental analysis
- Can be used by anyone with minimal computer experience
- A great learning tool for schools
- Assigns environmental points which you can grasp and **set goals** with
- Annual updates

"Ingenious but inexpensive, *EnviroAccount* will analyze your lifestyle in excruciating detail..."
John Fried *The Philadelphia Inquirer*

Other Books from Morning Sun Press

***Cooking with the Sun*: How to Build and Use Solar Cookers** by Beth and Dan Halacy, pb. $7.95

> "*Cooking with the Sun* is the best guide to building simple solar ovens and solar hot plates."
> *Buzzworm* Magazine

***The Fuel Savers*: A Kit of Solar Ideas for your Home, Apartment or Business** by Bruce Anderson, paperback, $4.95

> "Solar Energy will surely be the key to a safer future, and everyone can start entering the future now, thanks to this user-friendly, helpful book."
> Paul H. Ehrlich, Author *The Poplulation Bomb*

Order Form

To order EnviroAccount Software, before Feb. 1, 1994 make check out to EnviroAccount Software for $49.95 and mail to:

EnviroAccount Software
605 Sunset Court
Davis, CA 95616

After Feb. 1, 1994, please order by phone by calling 1-800-554-0317.

Please print the following ship-to information:

Name ______

Address ______

City/State ______ Zip ______

IBM Disk size: 5-1/4" ______ 3-1/2" ______

Macintosh: ______

To order books from Morning Sun Press:

____ copies of *The Fuel Savers* @ $.4.95 (U.S.)
____ copies of *Cooking with the Sun* @ $7.95 (U.S.)
____ copies of *EarthScore* @ $8.95 (U.S.)

Shipping: Please add $1.75 for the first book, $.50 per additional books. Allow 2 weeks for delivery. California residents add appropriate sales tax. Quantity discounts available. Please send check or money order to:

Morning Sun Press
P.O. Box 413
Lafayette, CA 94549
(510) 932-1383

Please print the following ship-to information:

Name ______

Address ______

City/State ______ Zip ______

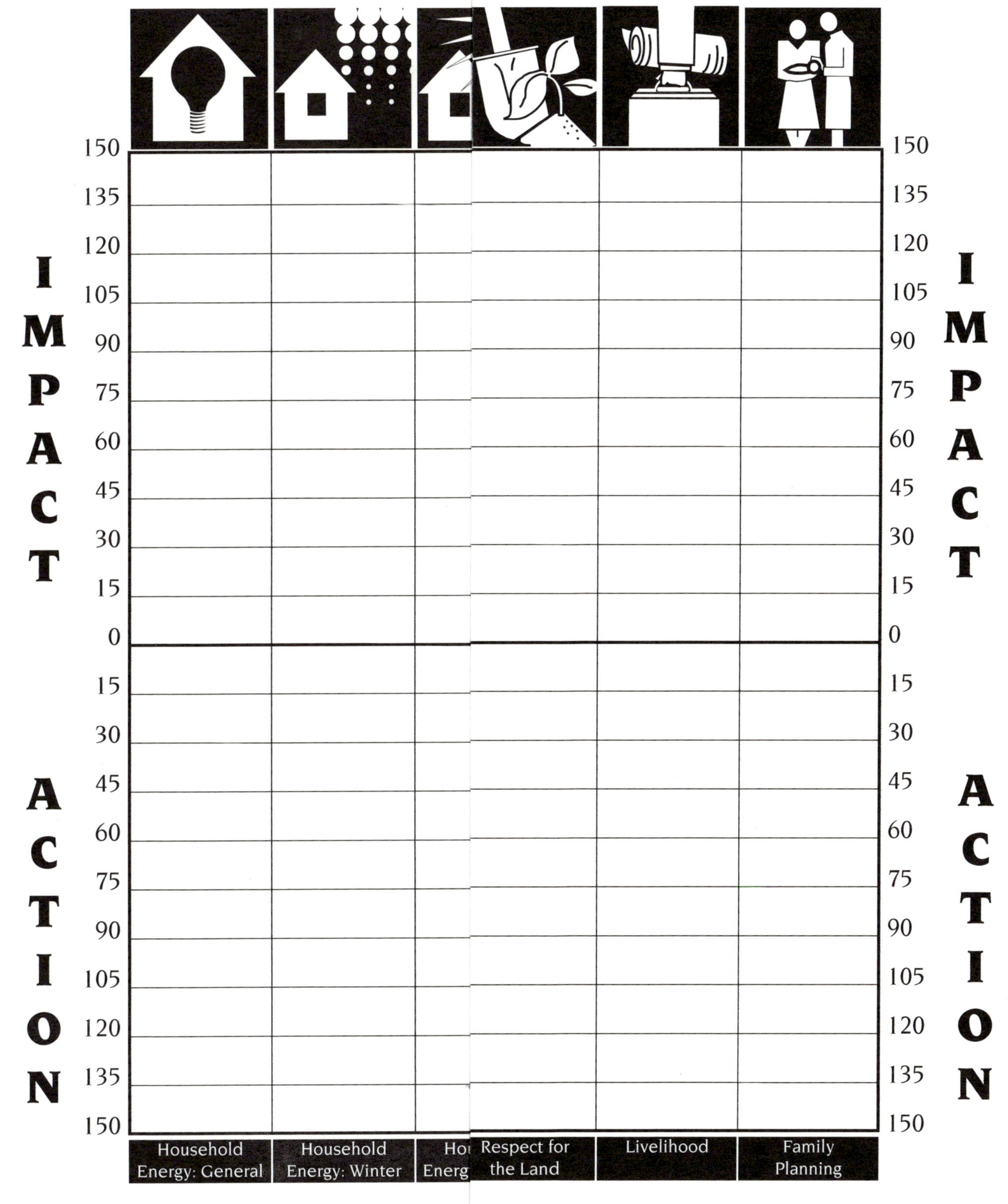

Impact Rating

Eco-Titan

Eco-Hero

Eco-Mentor

Eco-Average Citzen

I

150

TOTAL IMPACT POINTS:

ACTION RATING:

re Chart

aper and est Products	Toxics	Waste and Recycling	Envirnmental Advocacy	Respect for the Land	Livelihood	Family Planning	
							150
							135
							120
							105
							90
							75
							60
							45
							30
							15
							0
							15
							30
							45
							60
							75
							90
							105
							120
							135
							150

IMPACT

ACTION

tion Rating	Points
esome (Earth Action Hero)	more than 400
cellent	301-400
y Good	201-300
od	101-200
ould do better!	0-100

OTAL ACTION POINTS:

ACTION RATING: